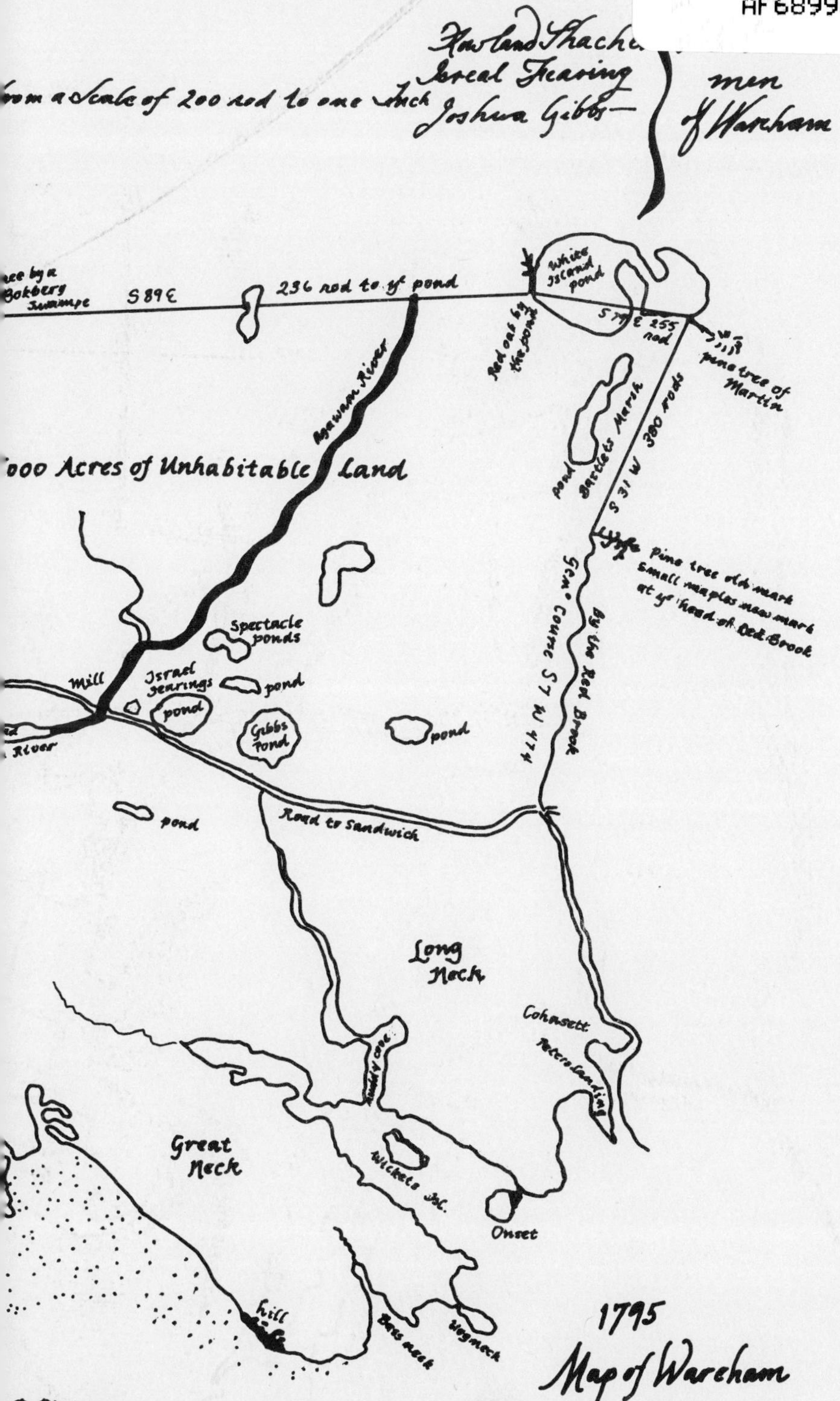
Howland Thacher
Isreal Fearing
Joshua Gibbs
men of Wareham
from a Scale of 200 rod to one Inch
S 89 E
236 rod to ye pond
White Island pond
S 77 E 255 rod
Red oak by the pond
pine tree of Martin
Agawam River
Bartlets Marsh
pond
S 31 W 380 rods
7000 Acres of Unhabitable Land
Pine tree old mark
Small maples new mark
at ye head of Red Brook
Spectacle ponds
Israel Fearings pond
pond
Mill
Gibbs Pond
pond
River
Gen. Course S 7 W 474
By the Red Brook
Road to Sandwich
pond
Long Neck
Cohasett
Peters Landing
Great Neck
Wickets Isl.
Onset
hill
Bird Island
1795
Map of Wareham

the Fearings and the
FEARING TAVERN
with the Bumpus Family

By Raymond A. Rider

Published For

The Wareham Historical Society

By

William S. Sullwold Publishing, Inc.

Taunton, Massachusetts

ISBN 0-88492-021-6

Printed in the United States

DEDICATION

This book is dedicated to

Irene Frances Rider,

my wife of fifty years, who first conceived the idea of restoring the crumbling old building, now known as Fearing Tavern, to its former pristine condition as a repository of Wareham's early history.

Without Irene's organizational ability and constant reminders of progressing programs, dates with all the contractors, and architects' committees, I could never have completed the work of restoration involved.

CONTENTS

FEARING TAVERN

PREFACE

The difficulty in identifying the early settlers, the Fearing and Bumpus families, was due to the lack of a middle name or initial, or more likely a careless recording in the town records. I have attempted to identify the individual by his wife's name or date of birth, his contemporaries, or some major event in his life or the times.

Therefore you will note some repetition of history to identify the individual. For instance, and to show the extreme difficulty of idenification, one Benjamin Fearing had two sons named Benjamin Fearing. Noted in the town records as Benjamin One and Benjamin Two, one of these Benjamins is later called Jr., presumably the first Benjamin.

Only an exhausting search of the wills *may* possibly simplify the problem. Accordingly, I have done the best I can do to identify the makers and shakers of Wareham's early history.

Records of these early times are sometimes found on microfilm, which when enlarged and combined with the blurring and early phonetic spelling are very difficult to read. Here too, middle names and initials are lacking, and the individual is identified only by the event, time, place and wife's name.

It is interesting to note that few of the Fearing families still reside in Wareham but many of the Bumpus families are residents, though they take little or no interest in town affairs, hold no offices, and are not noted for their financial achievements.

Their names, however, are prominent in all the wars as soldiers and sailors. They were noted throughout history for sharing the fighting drudgery imposed by the involvement of their country in wars of one kind or another. Many Bumpus men gave their lives for their country or made other great personal sacrifices as loyal citizens. The Bumpus name is honored on public works, such as a local bridge, for future generations to admire.

One also notices in a research of this kind how over the years land and other financial resources change hands from one or two families to another one or two prominent families, and how resources such as large land-tracts, mills and banking interests flow from one family name to another over generations. One notices how industries wax and wane, die and are replaced by others through ingenious methods inspired by new owners of land put to productive use as cranberry bogs, beach property, or shopping centers, and how a revived nail factory using ancient equipment continues to produce competitively in a modern world.

The growth of the Town of Wareham had also waxed and waned until the twentieth century. At that time the population began to increase by large numbers into a modern town government, spending millions of dollars in collected taxes for modern improvements and maintenance.

The Tremont Nail Co., now over 156 years in operation, is operated by Mr. James Kenyon and Mr. Shaw. They are very historically minded and have often made substantial contributions to the Wareham Historical Society. They have a continuous interest in the Historical District, next to which their business property lies. This company manufactures steel cut nails of the same pattern used over a hundred years ago, supplemented by additional types, both of which they ship all over the world.

The company operates a country store across the street from the mill in the old coopershop that Benjamin Fearing speaks of when he leased the property to "Isaac Pratt and Jarred Pratt and Charles Wilbur who improved the property."

A herring run or weirs (a ladder for the fish to get to the upper reaches of the pond to spawn) is still in use. The herring were of great concern to Israel Fearing and Jonathan Bumpus in the colonial period, as early records speak of "making way for the fish to go up and down the Wankinco river."

It was necessary for anyone who was a yeoman or a freeman and wanted to vote that he own property worth at least 20 pounds. The town meetings were never crowded, for that amount of money was possessed by only a few people, who made sure to keep control of the town government. The same names occur in the town offices over and over again. Bumpus, Swift, Fearing, Chubbuck, and Wilbur were sure to make an appearance as principal officials yearly. Squire Fearing was a Selectman, Treasurer, fence viewer and elected constable. Sometimes the Squire was elected as "town clark to sarve for one dallar for ye ensuing year."

The Benjamin Fearings served as town clerks. One Benjamin W. Fearing, an early settler in Rochester, was one of the first town clerks.

A section of this book is devoted to old taverns, inns, and ordinaries to show a comparison in structure, location and use. It was customary to insist when a license was granted that the inn should be close by the church. This was a great comfort after attending a heatless winter service in church. Benjamin's Inn was within sight of the early Congregational Meeting house as was "Parson Everett's Fulling mill," leased from Benjamin Fearing.

To understand these Pilgrim fathers as they are known, we must realize they came from an atmosphere of religious intolerance, and were Protestants in an overwhelming majority. Although they were very religious and almost everyone belonged to a church they were in turn very intolerant of the beliefs of other religious groups. The ruling element, clergy, merchants and civil representatives were repressive in demanding conformance to their views of Biblical interpretation. No work of any kind was to be performed on Sunday and church attendance was mandatory even in the extreme of the dead of winter in unheated churches.

Women and children were little more than slaves to the head of the household and a widow was a destitute and lonely creature until she could find another male to marry her. She owned no property until much later in colonial days. If her father was as wealthy a man as Israel Fearing, she got very little except the privilege to occupy a room in her brother's house and a few other bare necessities for living, which ceased upon her marriage. Sometimes a small sum of money was settled on her which became a dowry under the control of her husband. She was scarcely educated, since this was considered a luxury even in Wareham until the state of Massachusetts compelled communities to provide a minimum education for all children to be taught the 3R's and little else for generations. A school master (or mistress) was hired in Wareham to teach for three months in one section of the town, then a different section for each of the other three quarters of the year.

Some of the stronger female characters taught themselves from whatever books they could lay their hands on.

The movement to increase opportunities for learning began with Abigail Adams and continued without interruption up to the present time.

Abigail Adams, wife of John Quincy Adams (1767-1848), sixth president of the U.S., 1825-1829, was one of the most distinguished and influential first ladies in U.S. history to persuade her husband to "remember the ladies". Men such as Wright, Fuller, Mills and Martineau began their commitment to the cause for women first as theorists, then as writers.

Other women such as Anthony, Mott, Stanton and Stone took up the cry for more freedom and opportunities for women, which was finally transformed into a political movement. They continued to write and speak wherever the opportunity presented itself in the Nineteenth Century.

Great pressure and influence was placed on Massachusetts legislators and national leaders to listen to women, to vote for their proposals and finally to give them the franchise to vote for the eventual emancipation of women to more freely work for their legal rights established in law.

These New England settlers made their homes near the sea, up navigable rivers and in protected harbors, since the sea became not only a source of food but the only means of transportation for generations. They were dependent on the exchange of goods with foreign countries, which kept the shipbuilding industry active along the shores of Wareham from the site of Wankinco dam, down the river to and beyond what is now Pinehurst, then known as "Pine Neck."

The Fearings were active in this industry. When the British ship *Nimrod* anchored in the Wareham River in 1812 to send two hundred men ashore to destroy shipping, William Fearing's nearly completed ship was burned to the keel. When William Fearing complained to the British officer about this act of war he was rebuffed and lost his vessel for his pains.

Privateers were operated by Wareham men, the most famous being Capt. John Kendrick, whose home is now owned by the Wareham Historical Society and being restored to its former condition. Privateers were a constant threat to British supply ships bringing men and war materials to these Revolutionary shores. Many of these British ships were captured and their cargoes sold or confiscated by the Continental Armies.

CHAPTER I

New England People

The people of New England were unlike the settlers of Virginia, who were more pleasure seeking and less serious about life.

Massachusetts was settled by two colonies: the Plymouth Colony in 1620, and ten years afterward by the Colony of Massachusetts Bay. The first Colonists were known as the Pilgrim Fathers who in religion were called Brownists or Independents. They were as strict in their religious beliefs as they were in their daily habits. Pleasure was extremely limited. Their lives closely resembled the hardy climate of the area in which they lived. Long hours of grinding labor and poverty were the early lot of these dedicated people.

These Pilgrims were dissenters from the Church of England and had removed themselves from it, in consequence of which they were harshly persecuted.

The Massachusetts Bay people were Puritans and settled in the area now known as Boston Harbour. These two groups of Colonists were quite distinct in manners and beliefs, character and opinion.

Even after ten years the Plymouth colonists only numbered two hundred and fifty. At the end of seventy years they were absorbed by the Massachusetts Bay Colony and numbered only about nine thousand.

The Plymouth colonists were very poor in contrast to the Puritans of Boston Bay colony. These Puritans were a great power in England unafflicted by poverty and not compelled to hide or flee to Holland as the Pilgrims did. The Puritans counted among them some of the best educated, wealthiest, politically skilled, sturdy and virile people of all the English colonists. The English government then under Cromwell's influence was eager to see these people settle in America to offset the Dutch, who had a strong settlement at New York.

The next ten years of migration to America saw the population of the Puritans increase to over fifteen thousand, far outnumbering the Plymouth Colony, and begin to exert a commanding lead in government and religion.

The Puritans in England began to acquire under Cromwell a change in government and to remove the religious shackles they detested for so long to the extent that emigration ceased until 1840, when Irish and French Canadians came in large numbers to establish their own religious denomination.

Under the Puritans no man could become a freeman unless he was a member of a church. In 1670, when the population was twenty-five hundred, only eleven hundred freemen were counted. Most of the able bodied men had no more rights than the women and children. To become a freeman a religious examination of a very stern and rigid process was conducted to prove the earnestness and compliance of the candidate.

The minister of the church to which he was elected by the congregation exercised powerful influence to the extent that the governor and his assistants conferred with the ministers on the most difficult questions and took the minister's advice very seriously.

The minister held his power only as long as he was in office. When he ceased to be a minister of that church, he immediately became a layman with no more power than any other member of the congregation. In any event his power existed solely over his own congregation.

Under the charter of October 7, 1695, the Power of the Ministers gradually declined from 1695 until 1776 as real power was vested with the English-appointed governor and his assistants, who in turn ruled Massachusetts with an iron hand. Thus began the resistance to English rule culminating in the Revolutionary War and independence of the thirteen colonies aided by France.

The rule of England over the colonies came to an end. The era of the United States began and her increasing power has continued through two world wars up until the present day.

The decline of New England statesmen, poets and writers began with the inundation of the native population by massive immigration from Ireland, and southern European countries in particular. Many of the native New Englanders moved west to Michigan, Ohio and Illinois during the 1800's, where they were further diluted by the population already there. The Carolinas received some New England people.

Much of the early New England spirit and the Puritan ethic of hard work has disappeared with the "great inundation" of foreign immigration. Many early treasured buildings and architectural masterpieces have gone to the wrecker's ball.

The Society for the Preservation of New England Antiquities was organized and is bravely dedicated to prevent the further loss of the artifacts and buildings of early colonial days.

Wareham was a peaceful village for generations. The prosperity of the town depended on the sea for transportation. Many small schooner and fishing boats were built along the shores of Wareham. A shipyard was established by William Fearing "visible from his house" standing on land now owned by the Wareham Historical Society, part of which is occupied by the first Methodist Meeting House in process of restoration.

William Fearing owned and operated a store in the downtown area of Wareham supplying many of the necessities of life. It was located on land formerly occupied by the Old Colony freight station, now a parking lot. At this general location another Fearing Inn was in operation. Galligan's "Agawam Inn" flourished at this time also, across the street from William Fearing's enterprises

At the time Isaac Bump's house was built in 1690, Rev. Ichabod Wiswell and others went to England to get a new charter for the colonies, which was refused and a new one promised.

This promised charter on October 7, 1695, granted by William and Mary, united the colonies of the Massachusetts Bay, the province of Maine, Arcadia and New Plymouth including the Cape, into one province called the Province of the Massachusetts Bay in New England (noted in land deeds recorded in Plymouth, Mass.)

The Crown reserved the right to appoint the Governor, Lt. Governor and Secretary, retaining a firm control over the colonies. This right reserved by the Crown existed until 1789 (although ignored before that time) when the Federal Constitution became the supreme law of the land, vesting all powers in the people and annulling all charters. Thus began the squabbles of the 13 colonies over taxes, states' rights and many more individual prerogatives. The Civil War resolved most of these problems, leaving the larger question of Slavery and its aftermath unresolved.

The citizens of this Federation were always a restless and belligerent group, surrounded by wars between England and France, and preyed upon by both sides until war was declared upon England again in 1812.

The Pilgrims and the Puritans were remarkable people. They were constantly threatened by nature, the Indians and the Mother Country. They had their young males impressed in His Majesty's Service, shanghaied from water-front cafes, taverns and ordinaries. They were taxed into poverty, had their dwellings burned by Indians, and were slaughtered, tortured and scalped.

The Indians burned their corn and other grain fields as they matured. The Colonists struggled on farms to make a bare existence year after year with the whole family working from sun up to sun down. Women bore large broods of children, hoping enough would survive to take care of them in their old age.

Widows found themselves helpless and forced to work for a more prosperous farmer as a family servant or marry the first man who came along. They had no more rights on the whole than the children they bore. Property the husband owned was sold to pay his debts and to pay to the towns for support of the widow and her scattered children.

These times were grievously difficult, yet they survived to bring forth the greatest country from the Atlantic Ocean to the Pacific and from Canada to Mexico, the United States of America.

CHAPTER II

The Fearing Tavern and Others

The Isaac Bumpus lot, then 26½ acres, was reduced by flooding for a holding pool, a dam, a road and sale of part of the land to Tremont Nail Co. The lot on which the Fearing Tavern now sits is a part of the original Isaac Bump lot but has shrunk in size to about 2500 square feet. Part of the original lot is occupied by a gas station and a large diner. Another small building is now converted into a store and recently purchased by the Tremont Nail Co. for product display and sales. Another home and garage used as an antique shop occupies yet another portion of the original lot. The Fearing Tavern, now reconstructed, stands as proudly as before when it was a place to stop for the weary traveler.

A stage coach traveler coming down the Boston Post Road would pass directly in front of the Tavern on the south side, now known as Elm Street. Although the stately elms no longer exist, having fallen to the Dutch Elm disease as have most of the beautiful New England elms, two tall conifers towered over the entrance for years but were finally uprooted by the 1938 hurricane. The area looks rather bare with a highway barrier steel fence encompassing the adjoining parking lot. The Historical Society has planted a few slow growing trees on the south side and has an herb garden started on the east and north sides. A vacant lot adjoins the property to the east belonging to the diner. The Historical Society has plans to landscape this area later. The Society has acquired more land (ten feet extension) to the west in exchange for an undisturbed passage in the rear of the tavern on the north side where the diner is located. The building has a good setting now, looking from the south and front side, and is well kept.

The public is invited for a small fee three days a week.

The Society has five ther interesting buildings open only by special request of groups or individuals.

The Ordinary or Inn could accommodate more people than could be expected by counting the rooms, as it was usual to have a stranger share the bed with the present occupant, or if no more people could get into the bed, the stranger would be directed to sleep upon the floor on a straw mattress. Women rarely traveled alone, so there was no problem about mixing the sexes.

The Fearing Tavern drew visitors from afar only on horseback, as the roads were mere paths at this time, impassable to a stage coach transit. Only just before the time of the railroad did the stage coach stop at the Fearing Tavern for refreshment and a night's rest for its passengers.

The lot of land on which the Fearing Tavern sat originally consisted of 26 acres, but with flooding of the lower area near the river to make the mill pond for water power, half of the land disappeared under water and remains so to this day. Some of the land was lost to roads, Route 28 (Cranberry Highway) on the north side area and Elm Street on the south side, Tihonet Road on the west side and some private property on the east side, all of which reduced the original area drastically. Further division was made when a diner was set up on the open land to the north side and a division line permanently established when the Wareham Historical Society Inc. purchased the Tavern from Mr. Ernest Blanchard and Clara Blanchard for historical preservation and restoration. A pocket handkerchief 2500 square feet of land is all that remains of the original property.

Across Elm Street and to the east the Tremont Nail Co. operates, producing steel-cut nails in the same manner as performed one hundred fifty-six years ago. Mr. James Kenyon is the owner-operator of this successful business.

This nail manufacturing building has now been recorded in the National Register for Historical Buildings, giving it a special distinction and character.

The Wareham Historical Society persuaded the Town of Wareham to set up an historical district for further emphasis and preservation of its history. In the now established area lies the Fearing Tavern and the Old Methodist Meeting House, owned by the Historical Society, the old Town Office Building (town property) and the Congregational Church bordering the Center Park on which a memorial granite shaft bearing the names of soldiers and sailors who served in the country's wars are stamped on four huge bronze plates fastended to the granite. Two Civil War cannon point their snouts easterly on each side of the monument.

The Fearing Tavern has priceless and irreplaceable Fearing family furniture and other items of antiquity given, purchased and loaned to and by the Wareham Historical Society Inc. The public is invited to view these objects and enjoy a trip back into the Revolutionary War period and beyond.

The little old dwelling, seen from the inside of the Fearing Tavern, was built around 1690. Isaac who owned it was a grandson of Edward Bompasse. Isaac Bump was born March 31, 1642.

The Bumps are related to John Alden and his wife Priscilla through Alexander Standish and his wife Sarah (Alden), and through their daughter Lydia who married Isaac Sampson, son of Abraham, who married Bathsheba Matilda Bump.

The Bumps are also related to Mrs. Tom Thumb of Middleboro. She was a Bump who married (for the second time) Count Magri. Her maiden name was Mercy Lavinia Warren Bump. They are also related in a very tenuous line to Seth Pope and the Tobeys from Sandwich.

Israel Fearing bought from Isaac Bump the miller, this original dwelling with its land on the mill pond, containing about three and one-half acres. Israel owned the dam holding pond on the south side. The road going over the dam also belonged to Israel. Benjamin inherited and leased this land, the dam and water rights to "Isaac Pratt, Esq., Jared Pratt and Charles Wilbur, gentlemen, for the sum of three hundred and seventy dollars and ninety cents each succeeding Nov."

This Benjamin Fearing was the son of Israel Fearing born 21 Sept., 1644 whose ancestor was John Fearing who arrived in Hingham from England 1638 and died 14 May, 1665. His wife was Margaret. They had a son Israel born Sept. 1644, died June 1693 whose son Israel Fearing was born in Hingham 29 August, 1682 and married Martha Gibbs, died 1754. Their son, the first known Benjamin, inherited the Isaac Bump place. He married Salome Pope.

Sometime after William S. Fearing was born Dec. 26, 1804, Benjamin Fearing 2 built a home for himself near the Congregational Church and on the south side of the square at Parker Mills now called Center Park.

His son, Dr. Benjamin Fearing, inherited the property. Dr. Benjamin married Mary Fayette Mackie, daughter of Dr. Mackie.

References made to buildings and their locations can be seen on a map of 1879. Wareham's population in 1875 was 2,874, and the town was valued at $1,124,248.00. During Wareham's early history a war with the Indians and the Mother country of England drained the financial and human resources of the town. Wareham never failed to supply its share of men and material for the successful promotion and consummation of these wars from 1637 until 1677. As taxes from the King's representatives began to bear down on the colony, Wareham found itself once more caught up in a war. "Taxation without representation" circulated through the colony, resulting in the Revolutionary War, 1775-1778.

Seth Pope 3 was one of the original owners of the Agawam Purchase. Thomas Pope was his father. Seth married Deborah.

John Pope 4 married Elizabeth Bourne daughter of Elisha and Patience (Skiff) Bourne. Patience was the daughter of James Skiff, Esq. Elisha Bourne was the son of Richard and Bathsheba (Hallet) Bourne.

Sarah Pope 5 married Zaccheus Tobey, son of Samuel and Abiah (Fish Tobey, grandson of Thomas) and Martha (Knott) Tobey and great grandson of George and Martha Knott.

Tobey — the earliest record of a Tobey or Toby in our American annals is that of one Thomas of Boston.

The first known and credited ancestor of this family was Thomas Tobey of Scituate. He moved to Sandwich, Mass, and was a member of the church there in 1694. He married at Sandwich, Nov. 18, 1650, Martha Knott, daughter of George Knott deceased; children Thomas born Dec. 8, 1651 and perhaps others.

Samuel Tobey, son of Thomas, died Sept. 22, 1737. His wife's name was Abia and their children were:

II Joanna born May 22, 1697, who married Benjamin Spooner of Dartmouth, Nov. 29, 1716; Cornelius born Sept. 12, 1699; Tabitha born Nov. 9, 1701 who married Joseph Freeman May 9, 1726; Zaccheus born Jan. 14, 1703-04; Ruth born Sept. 8, 1706; Jonathan born May 13, 1709; Eliakim born Oct. 19, 1711; Samuel born May 8, 1715; Thomas born Aug. 14, 1720; and Elisha born 1723.

III Joanthan Tobey son of Samuel born May 13, 1709, married December 7, 1732 Abigail Lewis. His descendants are found in Dartmouth.

IV Jonathan Tobey, Jr., son of Jonathan born in 1740 married Dec. 28, 1777 in Dartmouth to Love White; son John 5 married Fannie Rounsville of Rochester.

VI Jonathan Tobey, son of Jonathan, Jr. and Love White born July 12, 1787, married Hannah Sears of Rochester.

Following is the beginning of the first records ordered by law, and kept in the Wareham Town Clerk's Office.

"The Records of all the proceedings in town meetings and all other records which by law are required to be kept for the Town of Wareham, containing first, records of the town officers, jury men, secondly, Births and Deaths of persons of the town; thirdly, records of the cattle marks of signatures; and fourth, records of establishments and managers within the town.

"All carefully Registered within this volume from the first incorporation of the Town of Wareham in the year 1739 . . ."

"Whereas Cap. Israel Fearing and Mr. Samuel Burge both of Wareham in the County of Plymouth in New England having land and meadow and enjoying it together, and being without sufficient fences, each party receiving damage for neglect of said fences being made and for this cause said Fearing and the said Samuel Burge in the month of March 1746 before Eldad Tupper Junr and Samuel Fearing Junr to make an maintain his half of the fence att Littleneck so called between sd Fearings meadow and lands and sd Burge his meadow and lands and also to make the part of fence between the first and second lotts in Wankinco lying below Jonathan Chubbuck."

"May 12, 1746 Capt. Israel Fearing warned Sam'll Burge again to make up his fence at Little Neck and also at Indian Neck before Israel Fearing Junr and Eleaszor King and thereupon Sam'll Burge neglecting to make up his part of the fence Cap. Israel Fearing applied himself unto us the fence viewers of sd town being chosen at the annual meeting in March 1746 and being sworn to the faithful discharge of the office we appeared on the nineteenth day of this instant May 1746 and proceeded as followeth: (VIZ) Beginning in the Indian Neck with a Pine Tree at the Cedar Fence as it now stands 72 Rods to a stake. This part for Cap. Israel Fearing to make and maintain allowing Sam'll Burge so many pine sticks as will make three rods and three feet of fence and that Burge shall cut sd sticks within a year from the date hereof. The other half part is to run sufficiently into the salt water as the fence now stands and this for said Burge to make and maintain and both parties, Fearing and Burge, their heirs and assigns are to make and maintain the fence as above mentioned as long as they improve the land and meadows as they now lye.

Another piece of land lying between the first and second lot in Wankinco said Fearing owning part of the second lot and Burge improving the first lot; their fence is divided as followeth . . ."

Fearing House, Parker Mills. Wareham, Mass.

CHAPTER III

Fearing Tavern Restoration

A local man who chooses to be anonymous gave large sums to assist the Wareham Historical Society in its reconstruction and restoration project. The Theodore H. Barth Foundation was generous also. In addition, many smaller contributions added to the sum needed to complete the job.

The president and founder (who is also the author) assisted by Mrs. Raymond A. Rider (who conceived the original idea of saving the old Fearing homestead, formerly the Fearing Tavern, sometimes known as Benjamin Fearing's Inn) labored for fifteen years to accomplish it, and has loved every minute of it.

The plan of operation was not the most practical from a builder's point of view, but from the Society's view it was the right method. By renovating the inside first we were able to arouse the community's interest and the interest of certain descendants of Benjamin Fearing who owned and operated Benjamin's Tavern.

The first concern was to keep the weather out of the house to prevent further damage. The windows had been broken, were rotted and missing entirely in some instances. That became our first project. To get the funds we memorialized them and raised $2,500 to have a copy of each decayed and rotted window made. The roof was temporarily repaired and the chimney flashings tightened.

The whole building was in a sad state of repair. To make a grand effort after the windows were installed, the society obtained a mortgage of $6500 to renovate four rooms on the first floor. Once the society was committed to these major repairs, the senior and junior Monday Club of Wareham wanted to participate in a memorialization of the Public Room in honor of the three Fearing women who lived in this house and were presidents in turn of the Club. These dedicated women lifted the spirits of the whole community by their glamorous and constructive programs as the Monday Club.

The Engels, Doctor Gilson Colby Engel and Mrs. Engel, who is a direct descendant, enlisted the support of the T. Rowe Price family and the Killhour family, the latter's distaff side also direct descendants of Benjamin Fearing. Well! Things began to hum and funds began to pour in from these families. The second floor bedrooms were memorialized and placques and pictures obtained until the work was completed.

In the Publik Room, where drinks were consumed and where the early selectmen met to discuss town affairs, and where grog was furnished at town expense while deciding the issues and John Fearing, esquire, sometimes held court, a plaque on the door panel on the west side of the fireplace memorializes former Fearing woman who were presidents of the Monday Club in Wareham. It states:

THE PUBLIK ROOM

Through the combined efforts of the Wareham Monday Club and the Wareham Junior Monday Club

In Memory of past presidents

Mrs. Angenette Fearing Shurtleff	1892 - 1893
Mrs. Delia Wing Fearing	1893 - 1894 - 1903 - 1906
Mrs. Mary Fearing Warr	1932 - 1934

A plaque on the door in the East Bedroom reads:

In Memory of Mary Fearing Goodall
Born in this house October 8, 1849
Given by her granddaughters
Doris G. Engel
Jean C. Killhour
Eleanor G. Price

These three daughters and their husbands contributed very generously to the Fearing Tavern restoration and cheered us on to the final completion of this enormous project.

A picture of Mary Fearing (1849-1946), married to William Bailey Goodall, is to be found hung on the South wall of the bedroom.

Another plaque, found on the hall side (east) of the door of the West bedroom, reads as follows:

In memory of Hannah Cushing Barrows
1806-1892
Mother of Mary Fearing

Now the last great project remained to be done. The great spreading enormous roof has to be entirely rebuilt. Over the centuries the roof timbers had been replaced in spots and reinforced in others, in a jumble of odd-sized and inadequate timbers. The main rafters, hand-hewn, were weak and rotted, infested with powder beetles and termites.

The shingles were asphalt — a modern product not in harmony with the ancient building, so we had decided on using shakes, a hand split wooden cedar shingle about two feet long of various widths and thicknesses, to replace the asphalt shingles.

We had difficulty in finding a contractor who was willing to take the necessary risks in removing the old roof for fear the side walls might spread and cause the collapse of the building. This fear was real but we found a skilled and clever man, Mr. William Maynard, who carefully removed one side of the roof at a time, securing the building as he progressed.

Our architect, Mr. Lloyd Hendrick, and the author, were constant watchdogs daily as the roof was taken down and as carefully replaced with new hand-hewn timbers. The purlins or intermediate horizontal timbers were mortised in as were the originals, then

Fearing Tavern Restoration

boarded up and down as was the original roof. Once this was done a 4 x 8 plywood panel was applied over the vertical boarding, a layer of waterproof paper applied over the plywood, and the shakes nailed on five inches to the weather. The roof is now as strong as it can be, supported part way up on the inside by "jacks" or upright beams mortised into the main beams and cross timbers tying the walls together. These jacks are original sticks mortised into the old roof rafters and the huge beams below, then pinned with oak pegs.

The entire roof construction followed the early style of mortise and tenon with oak pin construction. It looks good and is very sturdy. With care this roof should be in good condition for a hundred years or more.

We still had the old kitchen to finish, but our own funds were now beginning to accumulate sufficiently to renovate this ell back to its original condition. We had already shingled the roof as part of the main roof reconstruction. With that we had renovated the entire building, now open to the public Tuesday, Wednesday and Thursdays in July and August.

The building is presently unheated. Hopefully some day this can be adequately remedied.

We have a unique building constructed over three eras, completely furnished and not only a pride to the Wareham Historical Society but the town as well.

When Isaac Bump the miller owned the property it contained more than twenty-six and one-half acres. As time went on some of the land was sold to the Tobeys. A road was built over the land south of the pond area and just north of the Bump house, reducing the area to a little more than an acre. William H. Fearing, a subsequent heir, sold more of this property until little remained but that bounded by Tihonet Road, Route 28 and Elm Street. The easterly side was the property of the Tremont Nail Co.

This property never was a farm, as Isaac Bump was a miller owning three grist mills, one on Bump Brook and another on the Weweantic River. He undoubtedly shared in the sawmill enterprise with his father, John. The sawmill enterprise was beginning to wane from outside competition, and the timber was gradually being used up as more modern mills came into production. But the Bumps were involved in the iron industry as previously indicated, and thrived as a family until it began to fail. The Bump family began to sell their land holdings to others as their fortunes dwindled.

Restored front entrance

The Tobey and Makepeace families began to buy land, including bog areas. The Makepeace family became the dominant cranberry bog owners and land owners continuing up to modern times.

The great Tobey lands and industries have fallen to others but the Makepeace interests continue today.

The Cape Cod Cabinet Co. of Mattapoisett's own expertise was so outstanding they were given the project to manufacture the front entrance, drawn by Mr. Lloyd Hendrick, our gifted architect, from a contemporary entrance found in Kingston.

Because of the several entrances imposed on the building over the years we were not sure of just what the original one looked like, so we decided to look for one of that period conforming with the structural arrangement of the outside and inside of the dwelling. Some rooms, particularly the east and west second floor bedrooms, showed an ornate north wall paneled and moulded with fluted columns on each side of the fireplace indicating to us the excellent taste of the builder and owner of the first renovation. This addition was made in 1765 over and around the little old original two room structure, extending south and west and upward for a second story.

A large main roof covered the whole area but at this time was in a deplorable and dangerous condition. The Society shored the roof up with some temporary supports and repaired the leaks, one of which, aided by termites, nearly consumed the northeast corner post.

After completing the windows we borrowed from the National Bank of Wareham sixty-five hundred dollars and proceeded forthwith to renovate the first floor. Mr. John Rhodes was the contractor.

The Tap Room was a very important part of each tavern as that of the Sudbury Inn was vitally needed for dispensing liquid refreshment. A wooden grating (a fence-like structure) with a portcullis was raised or lowered according to the demand for liquor.

The Fearing Tavern has such an arrangement and can be seen restored with most of the original forms intact. The wide boards in the tap room are worn down into deep depressions between the knots in the wood.

A grill, without the gate in the center, was found in the attic in the northwest corner. It was brought down and fitted exactly under a decorated arched beam in the tap-room to the west of the Public Room. As history cited the fact that the selectmen were served "grog" at the expense of the town while conducting the town's business, we knew from this a bar must have existed in this old tavern and we found evidence enough to prove its location, and the grill fitted the space. The arched beam had never been disturbed all of these many years.

We also found evidence of an early "dumb waiter," a small elevator supported by ropes through a pulley with compensating weights, operated by pulling on the rope suspended from the ceiling on the second floor and running down the front and rear of the small elevator.

The floor above was apparently a storage for the liquor carried by way of the elevator. This was a small room kept securely locked.

In our restoration we had many hard decisions to make as to what should be restored, ignored or changed. In this instance we did not restore the "dumb waiter". We ignored the fact it existed. We turned the small room into a children's play room where all of the toys of the era are on display.

Mrs. John Cunningham donated twenty-five hundred dollars to restore a room on the east and second floor of the original "Bump

House" in memory of her husband, who came to Wareham for so many years and enjoyed the town so much.

Mrs. Stewart C. Woodworth, a sister to Mrs. Teresa Cunningham, wife of Dr. John Cunningham, was very interested in the reconstruction of the Dr. John Cunningham room, and she decided to donate the required amount to restore the adjoining west bedroom of the original Bump house and set a plaque in the memory of her sister, Mrs. Teresa Cunningham.

The idea spread and Dr. and Mrs. Engel with Mr. and Mrs. T. Rowe Price gave the funds necessary to restore the east and west main bedrooms in the first addition, being the two rooms on the front part of the second floor facing south, in memory of Mrs. Mary Fearing Goodall as noted.

There has been much written abount an underground railroad. The term underground is very confusing, especially to the young. It was thought to be a secret trail over which escaping Negro slaves from the south especially, were taken at night to way-stations set up to conceal them while in transit to Canada.

A legend persists that the Fearing Tavern was one such way station, but no proof of such a way station has ever been provided. It is unlikely a public place such as the Tavern would be considered, as too many people circulated in and out of the place. It is more reasonable to expect a way station to be on private property away from prying eyes.

The Lincoln Estate on Lincoln Hill has been established as one of the way stations for transporting Negro slaves northward. This is authenticated by the present owners of the property, who can point out the exact location of the hiding place.

It was suggested that a tunnel from the Tremont Nail factory to the Fearing Tavern was once open, but in the renovation of the Fearing Tavern no such tunnel or remnants of such a tunnel could be found. It is likely that the legend is merely a wishful hope without any foundation in fact.

A thorough search of every possible place of hiding was made from the cellar to the attic, behind fireplaces and walls, and no secret hiding places were found. The architect, the carpenters, and the author made a meticulous search as restoration work progressed. No place could be found that would indicate a passage way to a tunnel or other escape route to the outside.

In the east second floor bedroom is the beginning of the secret passageway where the two buildings of different eras are back to back with one closet leading into another. From there one can take a northerly direction thru a small door into this last addition we call the "Old Kitchen Area" and come down the back stairs unknown to anyone.

The legend is told of a British soldier hidden in the Tavern for several days while the local militia were frantically looking for him. Through the ease of changing locations by way of the secret passage it is said he was never discovered. It is a tale which never fails to make a lasting impression on visitors, and children are always fascinated by it. They can scarcely wait to go through the "secret passage."

Through the years the supporters of the modern Enrichment Program in the local public schools have made it possible for whole classes of pupils to visit and enjoy the Fearing Tavern. Many local organizations and other historical societies have visited our Fearing Tavern with great joy and enlightenment.

INTERIOR VIEW OF RECEPTION ROOM, SQUIRRELS' NEST.

CHAPTER IV

Inns and Taverns

The Fearing Tavern is representative of accommodations in Colonial and Revolutionary times. Some of these inns are still in existence and some are still serving the public, though now with modernized interiors, new furnishings, and modern plumbing.

Many books have been written specifically about taverns, one of recent origin by Irvin Haas, published by Arco Publishing Co., Inc. of New York called *Inns and Taverns.* The Fearing Tavern has a text and photograph therein showing its restoration and a sketch of its history.

Wareham had several Inns and Taverns. One of the most famous was the old Squirrel's Nest Inn, formerly Deacon Swifts Inn at East Wareham. Another Fearing Inn was on Main Street, then known as "Water Street" in the location where the old freight station used to be, now a parking lot beside the Fire Station on the east side and the Water Department workshed on the west side.

The Galligans had an inn and boarding house called "Agawam Inn" across from the Fire Station on Main Street.

All of these inns closed shortly after the Old Colony Railroad came into operation about the mid-1840's.

SOME STAGE COACH SCHEDULES
(This in 1836)

Duxbury every Monday, Wednesday and Friday 3:30 a.m. and evening Tuesday, Thursday and Saturday at 5 a.m. from Davenport Inn.

Newport and New Bedford from Boydens Tuesday, Thursday and Saturday at 4 a.m.

Providence mail from Davenport's every day except Sunday at 9 a.m. and 4 p.m.

Plymouth and Sandwich to Falmouth from Davenport's Tuesday, Thursday and Saturday at 5 a.m.

GROTON INN, GROTON

BULL DOG TAVERN, PROVIDENCE

ROGER MOWRY TAVERN, PROVIDENCE

Boston and New Bedford mail stage via Taunton leaves Marlboro Hotel, Boston, Tuesday, Thursday and Saturday at 7 a.m. Arrives Taunton at 1 p.m. and in New Bedford the same evening. Fare $3.00. "The Boston stage coach went no further than Groton (Mass.) in that direction" (northwest) and there could be found the Groton Inn.

Bulldog Tavern in Providence, Rhode Island, was one of several in that town. When the picture was taken it was as you see it, still serving liquor but under a different name. The building is showing signs of decay in the early 1900's.

ROGER MOWRY TAVERN

At Little Compton, R.I. a tavern was not only used as a meeting house, but a meeting house sometimes was turned into a tavern.

This is what happened to Mowry Tavern in Providence. The house which had been serving as a place of worship became a house of entertainment for travelers. The interchangeable character of these two public institutions gave the early settlers no shock.

It was often required that a tavern must be near a meeting house to acquire a license.

The Providence, Rhode Island inn opposite the state house

had a large beautiful golden ball hung over its entrance door to give meaning to its name Golden Ball Tavern. Its elegance as an inn drew the distinguished and rich to its doors. The most famous event to transpire at the inn was the Lafayette Ball in honor of Lafayette and several young Frenchmen accompanying him. When the evening arrived all the local beauties were on hand, elegant in rich flowered brocades over short quilted petticoats of silk or satin, with square cut bodices and powdered hair to set off, as did their high heel slippers with silver buckles, their exquisite woman's charm. The men were scarcely less gorgeous wearing silken hose and knee breeches adorned with silver buckles while rich brocaded coats, lace ruffles and powdered hair made them still more irresistible.

The "Gaspee," an armed British schooner acting to intercept traders, oyster boats and other private craft, and making forays ashore to pillage and take hostages, at one point became grounded as the tide receded. Because of the intense excitement due to the activities of Capt. Dedinston of the "Gaspee," a group of armed people congregated at Sabin's Inn on learning of the "Gaspee's" grounding while chasing the sloop Hannah commanded by Benjamin Lindsay off Providence, Rhode Island. Knowing that the Gaspee would be immovable for several hours, an expedition of local armed men boarded her and wounded her Captain, who was then taken ashore with the rest of the crew. The "Gaspee" was set afire and burned to the water line. The story was well known before the "American and Gazette" published the base facts. The British offered reward for news of the perpetrators but none was forthcoming.

David Arnold's Tavern was the scene of the succesful capture of a British Officer — General Prescott — by a band of forty men under the command of Major William Barton of Warren on the night of July 5, 1777. Permission was granted by Colonel Stanton, commanding officer of a regiment charged with protecting the populace from the depredations of the British soldiers who would come ashore and "indulge in the wildest sort of revelry." Barton was very familiar with Newport area and its islands. Prescott, as Major Barton knew, visited at night a house of a man named Orering about five miles above Newport. Prescott's errand was of no consequence to him and so the guard for him was light.

The forty men embarked in several boats to the island at Newport where they seized Prescott and brought him to Arnold's Tavern. He was fed and dispatched next morning to Providence

SABIN TAVERN, PROVIDENCE

from whence he was brought to General Washington's headquarters in New Jersey. The house dates back to 1765, and like the Fearing Tavern, town business was transacted there.

The Greenville Tavern dates back to 1730, a two-story building with open piazzas on each level for the merrymakers. Two huge chimneys penetrate the slanting roof, indicating many fireplaces inside the building used to heat the rooms. Rhode Island had many famous Taverns.

Some strange marriage customs were practised in Rhode Island as in all the New England States. One such recorded occurred as follows:

ELEAZER ARNOLD TAVERN, NEAR QUINSNICKET, LINCOLN

GREENVILLE TAVERN, SMITHFIELD

DAVID ARNOLD TAVERN, WARWICK

GOLDEN BALL TAVERN, PROVIDENCE

> ". . . on the 13th of September 1714 John Grant . . . did meet with Sarah Stephenson, widow, in the street within the town above said, stark naked save only her shift and they being lawfully published the said John Grant did accept in marriage the above said Sarah Stephenson, stark naked save only her shift, without housing or lands or any personal estate whatever, and in said street I did join together in marriage the above said John Grant and Sarah Stevenson on the day and year above said as witness my hand and seal.
>
> Nath'l Sheffield asst."

This was the method used by widows of debt ridden deceased husbands to cancel the obligations, a sort of bankruptcy declaration.

Various owners of Knapp's Tavern appear since the Revolution, though Margaret Knapp, daughter of the old tavern-keeper, held title to it until 1812. The Tracey family became the owners until 1901, over fifty years. Helen Reddington Adams, wife of Colonel Hershel Adams, bought the place late in 1901. Helen was a descendant of John Reddington, a Greenwich soldier of the Revolution, and later turned it over to the Putnam Hill D.A.R. as their headquarters. The house is tastefully furnished with period furniture of the Revolution. A desk in the house is said to have been used as the property of Colonel Barrett who led the troops at Concord. Upon this desk, Putnam, while a guest of Colonel Barrett, wrote a letter to Washington.

The first stage coaches were very crude wagons with access under the drivers seat to backless benches. There was no special place for luggage and the coaches were crowded and uncomfortable. Later the Concord Coach was put into service, a much more luxurious equipage compared with those then in use. The Concord Coach was first built in Concord, New Hampshire and brought greater comfort and speed to its passengers. When Israel Hatch put on daily coaches from Boston to Providence, about 1793, he covered the distance between 5 a.m. and 2 p.m., changing horses midway in Walpole.

To accommodate the more aristocratic of the Albany passengers, the famous Wadsworth Inn was built in Hartford, which still survives (1907).

Duxbury, like many old towns, had several interesting old ordinaries. In 1660 Mr. Collier was granted permission to sell liquor to his neighbors. He was one of several, one other being Constant Southworth, a Deputy. In 1668 Mr. Seabury was granted

KNAPP TAVERN, GREENWICH

INTERIOR OF KNAPP TAVERN (NOW PUTNAM COTTAGE)

OLD CONCORD COACH

WADSWORTH INN, HARTFORD

A CORNER OF THE TAP - ROOM OF THE WADSWORTH INN, HARTFORD

a permit to sell to his sober-minded neighbors in gallon measures only. The Puritan Ordinary was often the place where court was held.

The story of Madam Sarah Knight, who set out alone on horseback for the arduous journey from Boston to New York two centuries ago, was a very unusual experience for a woman alone in those early times. Part of the way she was accompanied by the postman, and when he was not available she hired an escort. The accommodations in the ordinaries were very crude and the roads rough and wild. It is believed that Madam Knight followed the road to New York on which the Vade Mecum for America issued in 1732 gives the names of all the taverns along the way. It is believed Madam Knight made her first stop at the Dedham Ordinary after leaving Boston. The first license for this tavern was issued in 1658.

The Liberty Tree Tavern is directly connected to the Boston Massacre on March 5, 1770. It stood on the east side of Washington Street, between Essex and Beach Streets. The book of Possessions says that in 1635 this plot of ground was given to Garrett Bourne who became a Freeman, built his house and planted several varieties of shade trees. In 1646 he transplanted an elm selected for its shape and vigor a little distance northeast of his house.

In about a century the house became a noted tavern, and a little later on the meeting place of the Sons of Liberty. Soon the transplanted elm became known as the Liberty Tree under which the Sons of Liberty rallied. On the 14th of August an effigy of the stamp officer, Andrew Oliver, was hung from one of its stout branches. The sheriff was ordered to remove it but the populace

OLD ORDINARY AT DUXBURY

WRIGHT TAVERN, CONCORD

There is no readily available information on these two taverns.

COOPER'S TAVERN, ARLINGTON

WAYSIDE INN, SUDBURY

TAP-ROOM, WAYSIDE INN, SUDBURY

LIBERTY TREE TAVERN, BOSTON

THE GREEN DRAGON, BOSTON

was so aroused that he refused. Some Tories, encouraged by their British allies and led on by one Job Williams, chopped the Liberty Tree down. At the close of the Revolutionary War a pole was erected on the stump, which served as a point of direction. The modern brick building now on this site (1907) has imbedded in its front wall a tablet with a Liberty Tree in bas relief.

At the Green Dragon Inn, Boston, Paul Revere, John Adams and others congregated. "In the fall of 1774," says Revere, "I was one of upwards of thirty, chiefly mechanics, who formed the members as a committee to watch the Tories and the movements of the British soldiers to gain military information. We held our meetings at the Green Dragon Tavern." The men of the Green Dragon gave the alarm on the eve of the battle of Lexington, spirited away cannon under General Gage's very nose, and in the course of the war did other dangerous work.

The Tavern stood on Union Street, and was a two-storied brick building with pitch roof, showing above its entrance a metal dragon.

One place Washington did stop overnight was the Williams Tavern at Marlboro. The place was still receiving visitors (1912) and welcoming guests. Erected in 1665 by Abraham Williams it has ever since been a popular resort for travelers. Old time cells

WILLIAMS TAVERN, MARLBOROUGH.

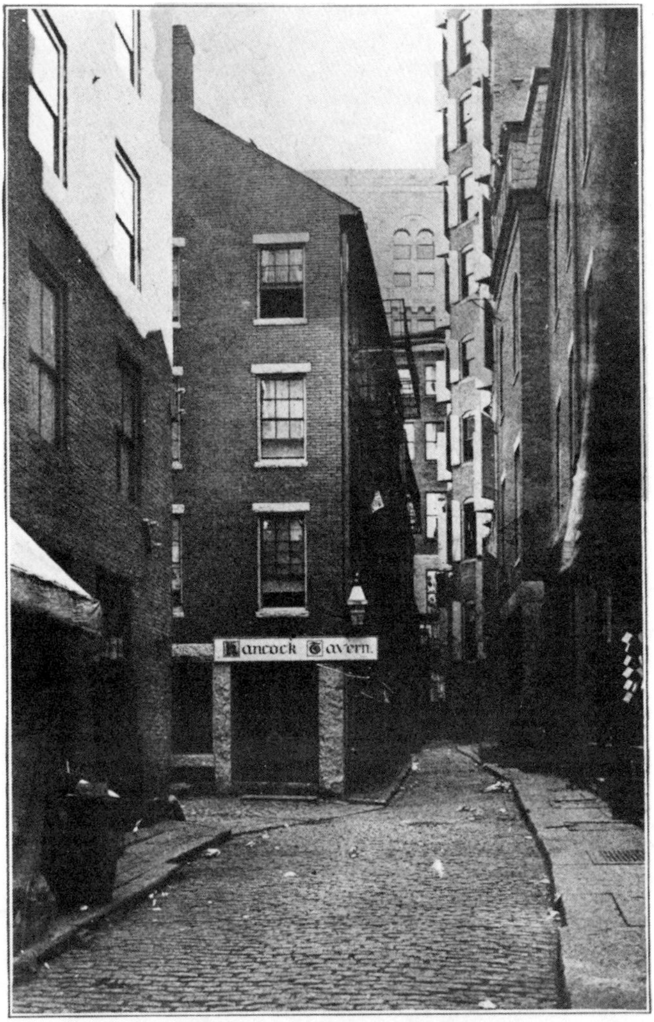

HANCOCK TAVERN, BOSTON

may still be seen, as court was often held there, it being the New England custom.

When Pease started his first line of mail coaches, this Williams Tavern was one of the principal stopping places on the mail route between Boston and New York.

An interesting story is connected with the Hancock Tavern and its owner in Corn Court. It is said Tallerand stayed here during his visit to Boston. The diplomat took a fancy to the inn-keeper's fancy pen knife and wanted to purchase it whereupon the landlord gave the fancy knife to Tallerand with the landlord's compliments. Later Tallerand became infatuated with a beautiful woman known to the world as Cordelia. The woman in her turn admired the knife so much that her lover gave it to her.

Later he deserted her. She was found dead on the floor of her apartment with an open note addressed to M. d. Tallerand on the table by her side. "I have burned all your letters," the note read, "They do no honor to my memory nor to your heart. You are the author of my death; may God forgive you as I do."

She had stabbed herself through her heart with the pen knife which had once been the property of a landlord of the Hancock Tavern.

Some of the Tavern owners were Tories and assisted the British whenever they could. Such was the case of Captain Jones, keeper of the Golden Ball Tavern, in whose house a British spy, John Howe, was hidden while out on a mission for General Gage, previous to the battles of Concord and Lexington.

Howe had stopped at another tavern earlier and by his questions aroused the suspicions of the crowd in the tavern. Howe left and walked toward the Golden Ball Tavern where he received a better reception as Captain Jones, a Royalist, recognized Howe as

GOLDEN BALL TAVERN, WESTON

WIGGIN TAVERN, HOPKINTON

a Britisher after some short conversation. Soon the crowd at the Joel Smith Tavern milled around the tavern yard and then started for the Golden Ball. Jones, hearing of the mob on its way to his tavern, told his Negro slave to escort Howe to the home of Mr. Wheaton, another Royalist, who lived in a remote section of the town. There he spent the night while 30 frustrated men searched every inch of the Golden Ball Tavern.

Colonel Elisha Jones, an ardent Tory, built the Golden Ball Inn in 1751.

A very lively sign long hung before Benjamin Wiggin Tavern in Hopkinton, on which was painted a soldier and the excellent advice "Pay Today and Trust Tomorrow." Lafayette was a man well honored as he passed through New England after laying the

BRIGHAM'S TAVERN (NOW THE WESTBOROUGH HOTEL), WESTBOROUGH

PEASE TAVERN, SHREWSBURY

WINN HOUSE, WOBURN

PAXTON INN, PAXTON

cornerstone of Bunker Hill Monument. Lafayette held a public reception in front of the Wiggin Tavern. School was dismissed for this famous occasion, according to a diary entry by Miss Betsey P. Eaton, afterwards Mrs. Brockway, a local school mistress.

Very often a tavern would be famed far and wide for its particular variety of drink. Brigham's Tavern at Westboro used to prepare mulled wine in this perfectly irrestible manner: a quart of boiling hot Madeira, half a pint of boiling water and six eggs beaten to a froth, all sweetened and spiced.

Some New England Taverns were famous for their spruce, birch, and sassafras beer, boiled with quantities of roots and herbs, with pumpkins and apple parings, molasses or maple syrup.

PEASE TAVERN

This Shrewsbury Tavern was one of three well-known taverns in the town — Farrar's, Baldwin's and Howe's. Farrar's is better known as Pease Tavern for the reason that the father of the turnpike became its owner. General Washington stopped off at this tavern while on his way back to Boston in 1789.

John Adams is said to have stayed in this tavern in 1774 while on a journey to some of the circuit courts in Massachusetts.

When John Tripp and his wife put up at the Bowen Inn, Barrington, Rhode Island, he received the following bill:

	s d
To 1 Dinner	9
To Bread and Cheese	7
To Breakfast and dinner	1.3
To 1 Bowl Toddy	9
To Loging you and wife	6
To 1½ Bowl Toddy	1.1½
To ½ Mug Cyder	1½
To 1 Gill Brandy	5½
To Breakfast	9½
Mug Cyder	1½
To ½ Bowl Toddy	4½
Dinner	8.
To 15 lb. Tobacco at 6d	7.6
To ½ Bowl Toddy	4½
To ½ Mug Cyder	1½
To Supper	6

Caleb Lord House is said to have had an extraordinarily congenial inn-keeper and model wife making this inn the most outstanding and desirable lodging house by acclaim of its patrons. The

CALEB LORD HOUSE, IPSWICH

success of any business depends on the devotion the owner invests to make it succeed. An Ordinary or Tavern is equally so, since its business depends on a continuous patronage and repeat business. Some people have the skill and "touch", others do not. This Ipswich Tavern had an inn-keeper who instinctively knew how to make people happy, hence his success.

Whippet House was another well known Ordinary serving the pubic.

WHIPPLE HOUSE, IPSWICH

Ross Tavern was built in 1734 and was in 1907 situated on one of Ipswich's busiest streets, under the shadow of a magnificent elm. The town boasts also of an inn at which Washington was entertained, long known as the Swansey House.

The charming home of Mrs. Harriet Prescott Spofford, near Newburyport's picturesque chain bridge, was also once a tavern. It was then close to the public highway and its landlord, Ebenezer

ROSS TAVERN, IPSWICH

Pearson, therefore was not exempt from suspicion when Major Elizah B. Goodridge of Bangor, Maine told December 19, 1812 of having been assaulted and robbed of a large sum of money near its doors. When Daniel Webster took the case he made Goodridge so contradict himself that all persons suspected of participating in the crime were discharged.

In March 1811 the other landlords of Byfield protested against Boynton's Tavern, stating that while it had been established for some time they doubted its continuance was necessary. Somehow the tavern continued to prosper regardless of the competitor's complaints.

HOME OF MRS. HARRIETT PRESCOTT SPOFFORD,
NEWBURYPORT

BOYNTON TAVERN, ON THE NEWBURYPORT ROAD

The Wolfe Tavern was named after General Wolfe, a British officer who helped to win the French and Indian Wars against England. Wolfe was a hero to William Davenport, who bought the house in 1743. In 1762 he made extensive alterations and additions and opened it as a house of entertainment for travelers. His sign was a board painted with the features of General Wolfe, his hero. William Davenport had been a captain of a company that had gone to Quebec to fight with General Wolfe.

THE NOBLE LADS OF CANADA

In a research of this kind many strange items are found, among which a sheet of verses entitled "The Noble Lads of Canada."

It is representative of the type of stiff opposition given to the British by the Colonial militia who mustered at taverns and way places before marching off to war.

The first verse reads:

> Come all you British heroes I pray you lend an ear
> Draw up your British forces and then your volunteers
> We're going to fight the Yankee boys, by water and by land
> And we never will return until we conquer sword in hand
> We're the noble lads of Canada—come to arms boys come.

The fifteenth verse concludes:

> Oh! if we reach Quebec alive, I'll surely stay at home
> For Mc Anough's gained the victory, the Devil take
> Macomb
> I'd rather fight a thousand troops good as ever cross'd
> the seas
> Than fifty of those Yankee boys behind the stumps and
> trees.

There are seventeen verses in all but these two verses tell the whole story in a very few lines. The paper is ancient, soft and very much deteriorated from folding, unwrapping, unfolding and folding again. It was found in a batch of material given to the Wareham Historical Society.

This tavern at the corner of Threadneedle Alley and Fish Street (now State Street) soon became a popular place of resort. The land travel from Maine to Boston went through Newburyport.

Tristram Coffin's wife was a clever and intellectually superior woman who charged 3 pence a quart for ale (the legal price was 2 pence) which she claimed was superior. Charged for asking an illegal price, she secured a witness who testified she put 6 bushels of malt into the hogshead instead of the legal minimum of 4, and so was discharged. Good wife Dionis Coffin had worked out a profitable problem for her husband-inn-holder.

This progenitor of the Newbury Coffins was also the founder of the Nantucket family of the same name. His son, Tristram, Jr. fell heir by marriage to another tavern belonging to the widow of Henry Somerby who died in 1652 (Edmund Greenleaf's license had been transferred in 1650 to Henry Somerby his son-in-law).

We do not know what manner of sign hung outside the Fearing Tavern. It undoubtedly resembled those shown on page 46, typical of tavern signs in the seventeenth century.

As you can see, taverns were "houses of entertainment" and without the sign looked much like any "ordinary" house of the times. Only the interior with its tap-room and long table with

WOLFE TAVERN, NEWBURYPORT

RICE TAVERN, KITTERY, OPPOSITE PORTSMOUTH

COFFIN HOUSE, NEWBURY

SIGN OF THE BENJAMIN WIGGIN TAVERN, HOPKINTON

SIGN OF THE WOLFE TAVERN, NEWBURYPORT

PURCELL HOUSE, PORTSMOUTH

benches distinguished it from any other house interior.

A large fireplace in or adjacent to the tap-room was the only means of heating either a tavern or a private home.

Levi Pease, the intrepid and imaginative stage coach driver who showed others how money could be made by driving stage coach routes between towns and cities, was responsible more than any other man for the success of the Taverns. They were the way stations between the start and finish of the stage coach routes.

The Rice Tavern in Kittery was just such a convenient stopping place. It was also opposite Portsmouth, a bustling, busy port for ships up and down the coast. Not too much is known of this tavern's history.

The Purcell House in Portsmouth had as innkeeper, Mr. Purcell, who had been proprietor of the Bell Tavern, until 1867 when it was destroyed by fire. Purcell's widow kept the Portsmouth boarding house known as Purcell House with which John Paul Jones was associated.

This came about as a resolution in Congress on June 23, 1779 "that Robert Morris should . . . take measures to speedily launch and equip the vessel *America* then on the stocks at Portsmouth, N.H." On the 26th John Paul Jones was chosen to command the vessel but found the ship only half finished. He was forced to settle down for quite a stay until the work was completed and chose to stay at what is now known as the Lord house (1907).

From a stagecoach schedule of the period: "The Portsmouth *Flying Stage Coach* operates with six horses, each person to pay 13s.6d. to Boston and 4s.6d. to Newbury. Sets out from the sign of the Earl of Halifax, every Tuesday morning, goes thru' Newbury to Boston and will put up at good inns on the road. The Master of the Coach is available Saturday to Monday night at Mr. John Stavers, innholder, at the sign of the Earl of Halifax."

John Stavers was so successful that in 1765 he left his first inn on Queen Street for the place which was still standing in 1907. The new inn was completed in 1770 and the old sign transferred but trouble soon arose because of the innholder's Tory leanings and the town's Sons of Liberty, who caused the innkeeper to flee for his life to a friendly Tory's house who himself was to be troubled from the same Sons of Liberty. One romantic story in existence concerns Martha Milton and Governor Benning Wentworth's "great house looking out to sea", where he resided.

From Haverhill, Washington and his distinguished party journeyed to Abbott's Tavern in Andover where he met Mr. Phillips,

ABBOTT TAVERN, ANDOVER

THE EARL OF HALIFAX (STAVERS INN), PORTSMOUTH

President of the Massachusetts Senate, who accompanied Washington and his party to Lexington.

Tafts Inn at Uxbridge made such an impression on Washington he wrote a letter to the proprietor informing him of his pleasure. Washington had learned that the proprietor had named one of his sons after the President and one of his daughter's first names after Martha, the President's wife's family, and being very pleased with the proprietor's daughters Patty and Polly, Washington made each a gift requesting it be kept quiet but that each girl write to him about it.

If George Washington had stopped at every tavern he is alleged to have been in, he would never have gotten out of New England.

At Conkeys Tavern as at so many other taverns a scene of rebellion was enacted by men who had been imprisoned for debts they could never pay. Deprived of supporting their families and comfort of their friends, they had rebelled against the judges who had them imprisoned. Cap. Daniel Shay, an adventurous soul who lived within a half mile of the hostelry (built in 1758) found it convenient to develop a plan there for resisting what he considered to be the tyranny of the judges. Shay saw, as did many others, that the people had become very poor by the enormous assessments on them to help finance the Revolutionary War. Judge Ward, who received the brunt of the anger expressed by the armed mob, persuaded them to allow him to work out a compromise, which he did. The rebellion collapsed once the militia assembled and dispersed the armed mob.

In Fitchburg, Massachusetts there is another tavern strangely connected with Shay's Rebellion. In 1907 it was the private home of Mrs. S. A. Fairbanks, previously known as the Upton Tavern.

CONKEY'S TAVERN, PELHAM

One incident told is of British officers in pursuit of rebels who fled there for protection. The servants of the law were unable to find the men in their diligent search; they failed to inspect a crooked closet under the stairway in the front hall.

UPTON TAVERN, FITCHBURG

The Fowler Tavern was so named for its landlord Fowler. A romantic story is related that the landlord's wife was kissed by British General Burgoyne, who was returning a prisoner to Cambridge and stopped at the inn. It was further related that the landlord's daughter was kissed. Finally a local historian maintained it was not Burgoyne but a German General Riedesel.

The Tavern may be called the place of the unknown kiss.

A judicial body, on a junket to hear the grievances of the Indians who were being pressed by land speculators to vacate their buildings, stayed at a Deerfield house which was later turned into a tavern and in 1907 was still standing. It is called the Frary House, and was in 1907 owned by Miss C. Alice Baker, a descendant of Samson Frary whose name the place bears.

The older part of the Frary house was built in 1689, the other half being added in 1748, when the whole house was used as a tavern.

FRARY HOUSE, DEERFIELD

FOWLER TAVERN, WESTFIELD

Back in the earliest days of the house it was the home of very simple people in whose family a young servant girl lived, Sarah Smith. She was accused of murdering her illegitimate child, and although she pleaded not guilty, she was condemned to hang. Her ghost is said to have haunted the place regularly for 40 years or more.

This two and a half story Tavern is unique as an early day building for this reason. The double porches add to its attraction for large groups enjoying the pleasant evening breeze while drinking their grog.

Many New England Taverns were visited by George Washington on his tour of New England and almost every Inn he passed by on his trip makes some claim to his historic visit to enhance the reputation of the Inn in order to increase its patronage or its historic value. But one inn at which George Washington stopped got a poor recommendation for its inadequate fare and lack of a silver spoon. Before he would attempt to eat the poor soup he insisted the proprietor obtain from any source a real silver spoon with which to eat. After much scurrying around one was found. Washington then consumed the soup.

YE OLDE TAVERN, WEST BROOKFIELD

BOLTWOOD'S TAVERN (AFTERWARDS THE AMHERST HOUSE), AMHERST

Ever since 1757 there has been a tavern on this spot. The first in a series was kept by David Parsons and his son, Gideon Parsons. Joel Dickinson, Solomon Boltwood and Elijah Boltwood were the successive landlords. Elijah's registers recorded the names of many men of national significance. In front of the fire at night sat doctors of divinity, learned lawyers, and members of the General Court and Congress.

So gifted in his profession was Uncle Elijah, however, that people in all walks of life felt very comfortable under his roof of convenience.

The Tavern in its day was a yellow building with a tap-room across the entire front of the building.

The sign of the lion was suspended from a stout post, indicating by its presence an ordinary or house of public entertainment.

About 1821, when Amherst College was established, the wooden building was torn down and a handsome brick building took its place. In 1838 Harvey Rockwood became the proprietor and the Boltwood Tavern became the Amherst House at about the time the stage line was replaced by the railroad.

Within a stone's throw of the Eagle Tavern, Horace Greeley learned the printing trade, and very often, no doubt, spent an evening in its public room talking politics to the other lads of the town. The Taverns were where people went to meet the stage, get the latest news and their mail. The Fearing Tavern served such needs in the early eighteenth century as Wareham's first post office.

Bath, Maine's shipping city, points with pride to the Shepard Inn, an old mansion which is still in the family who made it famous. The first sight to greet the traveler was a small window in the front hall which served as a bar and through which were handed hot toddies, gin fizzes and many other delectable drinks. The greatest treasure in the house, then as today, is the wall paper of the north parlor. This was brought from Italy more than a century ago by Captain James Hall, a relative of the inn's original owner, and is in colored sheets about a yard square. One of the scenes depicts the Vatican of Rome.

The subject that distinguishes this particular tavern is that representatives of Maine and Massachusetts met in its front northeast room in 1820 to settle a dispute. After a fortnight of meetings the Maine delegation agreed to recommend to the State Legislature that Maine would give Massachusetts $180,000 for her part of the public lands in that State. Of this sum $30,000 was in Indian claims which Maine assumed and the remaining $150,000 was to be paid in forty years with interest at 5%.

EAGLE TAVERN, EAST POULTNEY

It was only after a long session, during which the board sat in several towns and cities in Massachusetts, that they met at Jameson's Tavern in Freeport and signed the final papers.

Events that happened during the years the Fearing Tavern dwelling was in existence cover a period between 1690 and 1970, almost three hundred years.

The first events of major importance were the King Philip Indian wars previous to Isaac Bump owning the house. His family and brothers were involved in those wars, his uncle Samuel being killed at Rehoboth or Seekonk in 1675 by the Indians.

The Colonists were never free of war or the influence of war. England claimed all of the territory earlier explored by her Nationals. The King granted large parcels of land to individuals and companies who organized emigrations by the shipload of people who desired to escape from the devastation of the Continental Thirty Years War, its battles, burnings and lootings. Both Protestants and Catholics engaged in it. The poor people suffered the greatest hardships, many becoming destitute. They were recruited by the grantees who furnished food, equipment, seed and land controlled by the Proprietory Company.

This series of wars lasted from 1618 to 1648. There was continued war between European countries until 1776 when the Col-

SHEPARD INN, BATH

JAMESON TAVERN, FREEPORT

onists had their own war — the Revolutionary War of the thirteen colonies.

It was a war that made Wareham and other small towns destitute. Some profited by the war but most of the poor farmers became poorer. Heavy tax levies were imposed to pay for the war and its devastation. Wareham lost its share of killed and wounded. The Bumpus families provided many men for the Revolutionary War. Several Captains of privateers came from Wareham and several ships for this enterprise were built on the shores of the Weweantic River.

The Fearing families were strongly entrenched in the local economy during the very earliest Colonial history, but curiously began to lose control when the iron industries commenced to dominate the economy.

Several of the Bumpus women married men prominent in the iron industry. This did not seem to redound to the influence of the Bumpus family, seeming more like an absorption as the influential Bumpus families began to fade, much as the Fearing families were already doing.

The Makepeace family, the Tobeys, and later the Decas families became the leading political and economic factors, along with other new industrial and mercantile names which came to the surface: for instance, the Cornwells, a Main Street merchant family, and the Alden Besse heirs, a banking concern.

Many Boston based families purchased large tracts of land along the shore: the Welds, Minots and their families became summer residents on a permanent basis, year after year.

Once the Old Colony Railroad became operative through Wareham, the whole town economy was changed and its pace quickened, never to be the same again.

CHAPTER V

Land of Sippican, Agawam and Tihonet

In the earliest days of Wareham the land area on the south side of the town was a part of Rochester Proprietary or "Sippican." The North side belonged to Plymouth. It was sold to a group of men, later to be listed in this chapter. A third part was called "Tihonet", and belonged to Carver-Plympton-Plymouth. These three sections of land were incorporated into the town of Wareham, first the lands of Rochester and the Agawam Purchase, then Tihonet much later. Some of the residents of Wareham voted against taking Tihonet lands for fear the town would be getting some paupers to support, but the vote was in "ye affirmative", and the owner of the iron mill on the Tihonet pond area had his wishes granted.

The Fearings and Bumps

The Bumpus family name is registered first on these shores in the person of Edward Bompasse who came to Plymouth Colony in 1621 on the ship *Fortune*. The name Bompasse became Bumpus, Bumpas and Bump, among other spellings.

These two families, Bumpus and Fearing, were to buy and sell land, work together in town affairs and have great influence on the traditions and mores of Wareham.

Both the Bumpus families and the Fearing families owned large tracts of land, the Bumpuses by purchase and the Fearings by grants and purchase. This was the basis of their wealth. Men in each family had great respect for each other.

The Fearings were more aristocratic. Israel became Esquire or Squire Fearing with political power based on the King of England's favors and expectation of control by his squires as his representatives.

The Bumpus family was a hard working lot. They owned several saw mills, cutting timber from their large holdings of land and converting the logs into lumber of all sizes, from which John and Jacob and their sons built many houses. Some are still standing, and will be noted in later chapters.

General Tom Thumb and His Wife

The Fearings married into the Bump family, the Seth Pope family, the Bournes and the Makepeaces. This early large and expanding family was related to the Tobeys, the Lincolns, the Nyes and the Swifts from Sandwich, which then bordered Wareham at the Cohasset Narrows (now Buzzards Bay). They were related by marriage to Parson Thacher, Dr. Perez Doggett, the Briggs family, the Gibbs family and many others.

The early Bumpus family had several very succesful and enterprising men. One very famous Bumpus girl became world-famous as Mrs. Tom Thumb. More on this episode later.

The interplay of forces between the Fearings and Bumpuses ultimately produced the Fearing Tavern. Benjamin Fearing was the first Tavern innkeeper upon inheriting a large estate from his father, Israel, who bought from Isaac "Bump the Miller" all his mills, water rights, lands and his "dwelling house". The house is preserved today much as it was when Isaac Bumpus owned it, 300 years ago.

It was fortunate that when Benjamin Fearing increased the size of the tavern he retained the original structure by building over and around this Colonial Cape. One can observe the original timbers and flooring, the old chimney and fireplaces. A new roof was constructed by the Wareham Historical Society, Inc., using hand hewn rafters the same size and length of the original beams.

Israel Fearing, born in Hingham September 1644, married Elizabeth Wilder, daughter of Edward and Elizabeth (Eames) Wilder. Their son, Israel Fearing, born in Hingham August 29, 1682, married Martha Gibbs, and moved to Wareham where he became

John Bumpus House
Fresh Meadow Village, Wareham

Squire Fearing. He inherited land granted to his ancestors, which was in the Agawam Purchase called the Six Great Lots. This was probably the greatest reason for his coming to Wareham, so that he could more closely supervise the properties.

The genealogy includes the descendants of Israel who became prominent in Wareham history. The texts explain their office and place in the community.

The Fearings were very active during this period. They were engaged in nearly every business enterprise from salt works to ship building, from mercantile enterprises to grist mills, saw mills, fulling mills and iron mills. They owned or controlled many streams on which they had erected dams for water power.

Shortly after 1848, when many of the iron mills had ceased to operate, Fearing families begin to disappear. Although the Fearings were large land owners and involved socially, politically and financially in the vital resources of Wareham, the control of industry was rapidly transferred to other hands and pocketbooks.

Many of the Fearing families went south and west to more fertile fields of investment and enterprise. Today all the lands formerly owned by this branch of the Fearing family have passed into other hands, sold by succeeding heirs in large measure to the cranberry bog owners, a gigantic and growing industry. Their lands, in turn, are slowly being sold for development of industrial and subdivision purposes.

The great forests of pine and oak have succumbed to the ravages of man and forest fires. The only great trees still standing are on an island in Long Pond where the fires could not reach them.

The land area on which the Fearing Tavern is located was not part of the Agawam Purchase, in which Israel Fearing had large acreage. This particular area was a part of the Plympton, Plymouth and Carver area west of great lot six of the Agawam Purchase, now called Tihonet.

This great tract of land had several owners of a divided interest. Among these were Isaac Bumpus, Chubbuck, and Besse, and later the Tobeys, Crockers and Ellises around 1819.

According to a legend related in a history of Plymouth County, Massachusetts, "Tihonet" was obtained from the name of an early settler "who had a small right in one of the lots of land on which he lived and by virtue of his claim exercised the right of ownership of the land around him, so far as to cut the best timber and market it." Another gentleman thought he would look at the timber in that region with a view of purchasing it. He took this settler and his guide to show him the extent of the country and after traveling all day and asking the question often "who owns this lot" he was always answered by his guide "I own it." He soon discovered the trick and gave the old man the name Ti-Own-it by which he was afterward called and that section with a slight alteration in the name has borne his name ever since. As noted, a family of Chubbucks and Besses who early settled on this tract were as poor as the land itself.

This large land area contained two thousand and twenty acres and forms a nearly equilateral triangle. The Wankinco River, fed by the Tihonet Brook, drains the whole area and because of a fall of twenty eight feet at Tihonet dam, provides enormous water power sufficient to operate several mills.

The Agawam Purchase

The Six Great Lots were bought from Plymouth to be sold to raise funds for the construction of a new meeting house. This land had previously been purchased from the Indians and the deed duly recorded.

The purchasers of the original Agawam Area were:

John Chubbuck	Joseph Bartlett
Samuel Bates	Josiah Lane
John Fearing	Ephraim Wilder
William Beal	Nathaniel Morton
Seth Pope	Joseph Warren

Some of these men were speculators who quickly sold to others for a profit. The share were not equally held but divided according to the investment.

A few years later we find these names on the plot of land:

Ebenezer Burges	David Bates
Gershom Gifford	Joseph Hersey
Thomas Tupper	Joseph Warren
Samuel Bates	John Gibbs
Isaac Wilders heirs	Jirey Swift
Timothy Brown	Oliver Norris
John Bourne	Adam Jones
Israel Fearing	

One of the first things the proprietors of the Agawam Purchase did after acquiring the land was to take equal shares in the marshes, meadows, swamps and upland, the forests and streams. The marshes, especially the salt marshes, were very important to the early settler for the natural production of grass, fiber and food.

Herring Ladder

From the Indians the Pilgrims learned to use the cat-tails for stuffing mattresses and the down to stuff pillows and quilts, as the Indians used the down from cat-tails to pad moccasins and papoose board. From the fiber they wove coarse mats and baskets. They made dolls and other toys.

Roots were eaten as vegetables, made into soup, syrup and topping for puddings. Porridge was made from the pollen. Some of the roots were stored for winter dressing.

Dressings for wounds the year around were made from the cat-tails. Cat-tail soup and jelly came from the roots of young growth, and the hollow stems were used as candle moulds.

The cat-tails, when dipped in tar and turpentine mixture, served as torches to light the way in the night.

The lush salt marsh hay was greatly valued, cut in large quantities and stacked right on the marshes for winter use as needed. A raised wooden pole structure was specially erected with metal pans on the legs and center pole to prevent rats and mice from wintering in the stacked grasses.

A special wide wooden shoe was applied to the horse's hoof to keep the animal from sinking into the soft marsh while working to gather the salt marsh hay.

Catching Herring on the Wankinco River, Wareham

Cows and horses need a certain amount of salt in their diet. The salt marsh hay served to provide salt as well as the excellent nutrients in the hay itself. Cutting salt marsh hay was as necessary to the well being of the horses and cattle as cutting fresh meadow hay for regular fodder.

The marshes also concealed enormous amounts of fiddler crabs and other crustaceans available the year around. Water fowl of many species were taken for food from the marshes during migrations.

The salt marshes were sort of a cupboard provided by nature with no care or feeding needed by man. All the nutrients came naturally down the feeder streams from the uplands. These nutrients supported an ecological system feeding the sea creatures continuously through the centuries, providing man with constantly expanding schools of fish and supporting another important industry, the shipbuilding business.

BIDS ON WICKETS ISLAND

James Burges	bid	1.50
Moses Fearing	bid	2.00 in 1800
Cap. Abraham Gibbs	bid	1.34 in 1801
Cap. Joseph Gibbs	bid	2.50 in 1802
Joshua Gibbs	bid	2.50 in 1803
Robert Besse	bid	3.50 in 1809
William Swift	bid	2.60 in 1809
Jonathan Gibbs	bid	2.83 in 1809
Thomas Pendleton	bid	2.00 in 1810
James Gibbs	bid	2.92 in 1811
James Gibbs	bid	9.90 in 1812
Thomas Warren	bid	9.83 in 1812
Prince Savary	bid	2.50 in 1814
Contra price of Eliza Swift for James Burges bid		1.50 in 1803

Israel Fearing gave land and provided a gate to and in the Burying place, ½ acre for the Bumpus family lying on the westerly part of said buying place in the south westerly corner, part of the condition being that Israel Fearing was not to maintain the gate.

A boundary agreement in 1746 is recorded in the following manner:

Beginning at a bound tree numbered as appears by record by said Chubbuck land so running one halfway to the other bound tree numbered as appeared by Record by and this is land Cap. Fearing part to make and maintain so long as they Improve as the land now lies, and P. Fearing shall allow Sam'll Burge for his old fence four shillings old ten, and said Burge is to make and maintain the other half part of sd fence to the oak tree, so long as they use or improve the lands as it now lies. Another piece of land and meadow adjoining between Capt. Israel Fearings and Sam'll Burge at Little Neck, Said Burgess is to make and maintain, beginning at the lower edge of the creek on between their meadows and running to a rock upon the upland and thence by the line as they have divided to an oak stake which they have set there for a bound mark on four sides.

And said Cap. Fearing is to make and maintain the other half part from the stake to the turn in the fence and along to the other Bound which they have made. And also said Burges is to make and maintain the other part to John Bumpus his line, the First Part and the lot part being one half for Sam'll Burge to make and maintain he or his heirs and asignes and the other half part for Capt. Israel Fearing his heirs and assigns to make and maintain between his homestead and Sam'll Burge his homestead from the road that leads down into their neck, Northward till it comes to said Burge his Bound, this for Cap. Fearing to make and maintain his heirs and assigns and said Sam'll Burge is to make and maintain the fence on this South side of Popes Road ot a white oak tree and then round Pope's Point as far as Cap. Israel Fearing meadow goes this to Sam'll Burges his heirs and assigns to make and maintain.

The above written is an equal division of the Fences to be made in Partnership between Israel Fearing and Mr. Samuell Burge and this according to the best (of our) judgement of the land and fence we were applied unto to view and hereunto Let our hand this 20th day of May Ainogus Domini one hundred seven and forty-six.

Record May ye 24th 1746
by me Jonathan Hunter Town Clerk

George White
Thomas Bates

At the house of Israil Fearing a meeting was held to determine the:

Proprietors of Herring River	Share parts	26 October 1773
Barnabas Bates	2	
Noah Fearing	2	
Israil Bates	2	
Joshua Gibbs	1	
Prince Burgess	1	
John Harris	1	
Israel Fearing	2	
Benjamin Burge	1	John Gibbs
John Bates	1	Jany Burges
David Saunders	1	Jonathan Gibbs
Consider Sturdavent	1	Samuel Burge
Jacib Nye	1	Eben Swift
Joseph Sturdavent	1	Joseph Bates
Benjamin Burge	1	Joseph Saunders
David Besse	1	Jabez Besse
Jabez Besse 2nd	1	Samuel Harris
Jabez Burge	1	
Eben Bourne	1	
John Fearing	2	Thomas Bates
Israel Fearing 2nd	1	

Fearing Activities (Proprietors' Records)

"October 18, 1773 Voted to extend Red Brook into White Island Pond by digging a ditch and making a ditch. Voted Isaac Bowle to be a proprietor in the Brook, and by vote Noah Fearing, Jonathan Gibbs, Ebenezer Burge, Israel Fearing and Barnabas Bates to be a Committee to view the ground, whether it is best in their judgement to go any further in making the brook. Voted the Committee should bring in their report to the proprietors at ye adjournment of this meeting."

On October 10, 1772, Israel Fearing bought a book to keep his records when George the First was King. The book tells of indentures and agreements, trades and barters, impressments into the King's military service, of marriages and trials by His Majesty's Justice of the Peace. It tells of the habits of the people how they lived, paid their debts and conducted their religious life. It tells of their contentment in the small savings and small gains of simple pleasures and of growing families. The book is now owned by the Wareham Historical Society.

Some marriages were called "smock" marriages. When a bride was an insolvent debtor and it was necessary to prove she was possessed of no goods whatever, she was married with no more clothes on than her shift and this fact was certified in the justice's records. It was by law necessary that the bride cross the road barefooted and bareheaded before witnesses to be married on the other side.

Israel Fearing was impressed himself into the King's military service "May the 26, 1707 I was pressed to the casel for 6 months".

Impressments were avoided by getting and paying a substitute or running away. "Sept. 1746 Jonathan Chubback was impressed for Zeccues Bump at 5 pounds and 10 shillings old tenner for Edward Bump to go in his Majty sarves".

"March 1748 Judah Swift and Joseph Doty and Edward Rayment Imprest and hired Robart Bese (Besse) for 55 pounds."

Squire Fearing's house was on Great Neck Road in East Wareham near the Agawam cemetery in "the neighborhood", so called, and just beyond Pope's pond going down the Neck Road near the Bates' house. This small dwelling has been burned down by vandals. The area is now occupied by a home constructed by Clarence Hall now owned by the Warren H. Saverys.

John Fearing, Esq., son of Israel from Hingham, had his house on "Fearing Hill" on Fearing Road, as it later became known, then known as County Road coming from the east and going down the western shore of the bay to Sandwich.

"The road is sandy and difficult to travel. Only tradesmen of one sort or another travel this road and sometimes the local doctor who carried a bag loaded with magical bottles of pills to relieve illnesses and ward off fevers."

John Fearing, Esq. was a witness to the Agawam Purchase with Josiah Lane and so noted the recording in the Proprietor's Book. At the same time he desired Nathaniel Beale to be "clark" as before.

In the list of inventory from Israel Fearing's estate under "sundry other things" John had a pair of Nee Buckels Silver, valued at 4£". Having taken himself a wife he now became the proprietor of the Farm on Fearing Hill; and having been appointed to succeed his father in the office of His Majesty's Justice of the Peace, the title "Squire" was appended to his name. He was held in great esteem by the local people. He had control of both the domestic and civil life of the Community — everyone spoke with great respect when in the presence of "the Squire" or suffered the consequences of a fine or imprisonment in the stocks, which were just outside brother Benjamin's Tavern, for that place was where the Squire held court to judge offenders. The whipping-post was also in sight of the meeting house on the grounds of Benjamin Fearing's Inn. The squire also held court in his own home, as his book records. (This book is now owned by the Wareham Historical Society).

A Benjamin Fearing in 1850/1878 owned a dwelling house, was a voter and paid a poll tax. From an assessor's book of records for 1878 we have the following:

> A Benjamin F. Fearing owned 1/4 of a dwelling house, Caroline Fearing owned 1/4, widow Hepsibah Fearing owned 1/3 of a dwelling house. Widow Kirzen owned a dwelling house. William S. Fearing owned two dwelling houses, 1 store, 2 barns and a grist mill, 66 acres of tilling land, 10 bushels of rye, 50 bushels of corn, 4 acres of upland mowing and orchard, 40 acres of salt marsh, 40 tons of hay, 40 acres of pasture land and orchards, 12 cows, 100 acres of woodland, 1744 acres of unimproved woodland, 104 acres of woodland, 1 horse, 2 cows 3 years old, 4 steers and heifer 1 year old, 1 swine, 6 months old and upward, 100£ value real estate not before enumerated. William S. Fearing 1850-1878.

This receipt indicates W. H. Fearing was town clerk, tax collector and treasurer of the Town of Wareham in 1888.

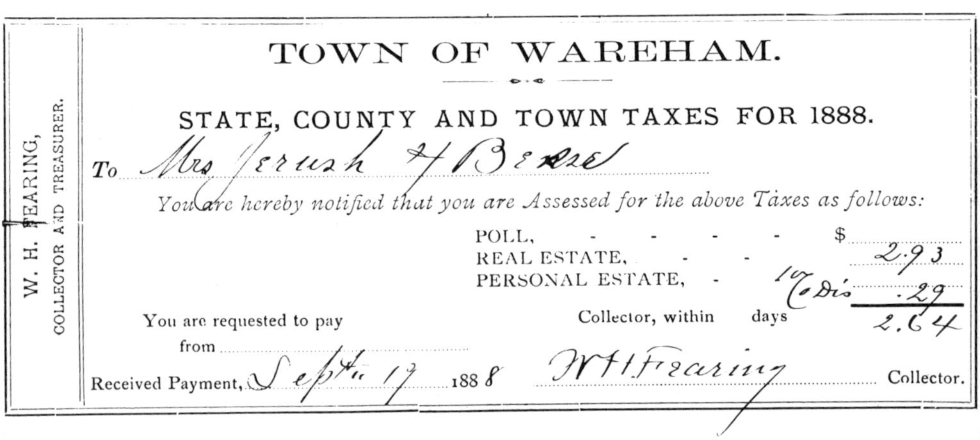

W. H. FEARING,
COLLECTOR AND TREASURER.

TOWN OF WAREHAM.

STATE, COUNTY AND TOWN TAXES FOR 1888.

To Mrs Jerusha H Besse

You are hereby notified that you are Assessed for the above Taxes as follows:

POLL, - - - - $
REAL ESTATE, - - 2.93
PERSONAL ESTATE, - 10% Dis .29
2.64

You are requested to pay Collector, within days from

Received Payment, Sept 17 1888 W H Fearing Collector.

CHAPTER VI

Reference in Old Rochester Records of The Proprietors and Place Names

1. Weweantic Pond —

A map drawn by Sylvanus Bourne in 1832 shows a pond area with a Weweantic Iron Works.

The same map shows a cotton factory on the Weweantic River near a bridge over the Weweantic River on County Road, now Fearing Hill Road. References to this cotton factory in William Root Bliss' *Colonial Days* places it near this location also, above this on the same river near Paper Mill Road. On the above 1832 map is found a Grist Mill, and further up the river another cotton factory.

A Grist Mill in Rochester at the Hiller Farm is shown on this map as being well up the Sippican River near the Wareham-Rochester line.

The old M. E. Church at the intersection of Carver Road and Tihonet Road is plainly evident.

A paper mill was located above the Weweantic Iron Works and the cotton factory.

A school house sits at the corner of Blackmore Pond Road and Fearing Hill Road, formerly Old County Road.

Wareham Iron Co.'s upper works at Tihonet, the lower works being below the locks (Tremont Nail Co.) area.

A salt works existed on the upper reaches of the Broad Marsh River (Pinehurst area), and another in the Hamilton Beach area. Little Harbor area supported a salt works as did the Swifts Beach area.

On the Agawam River in the same location as the Agawam Nail Co., stood a furnace, and a forge, was another Grist Mill.

This same 1832 map of Old Wareham shows the location of ponds and streams existing in times as early as Israil Fearing's era in the early seventeen hundreds. The place names, such as Pine

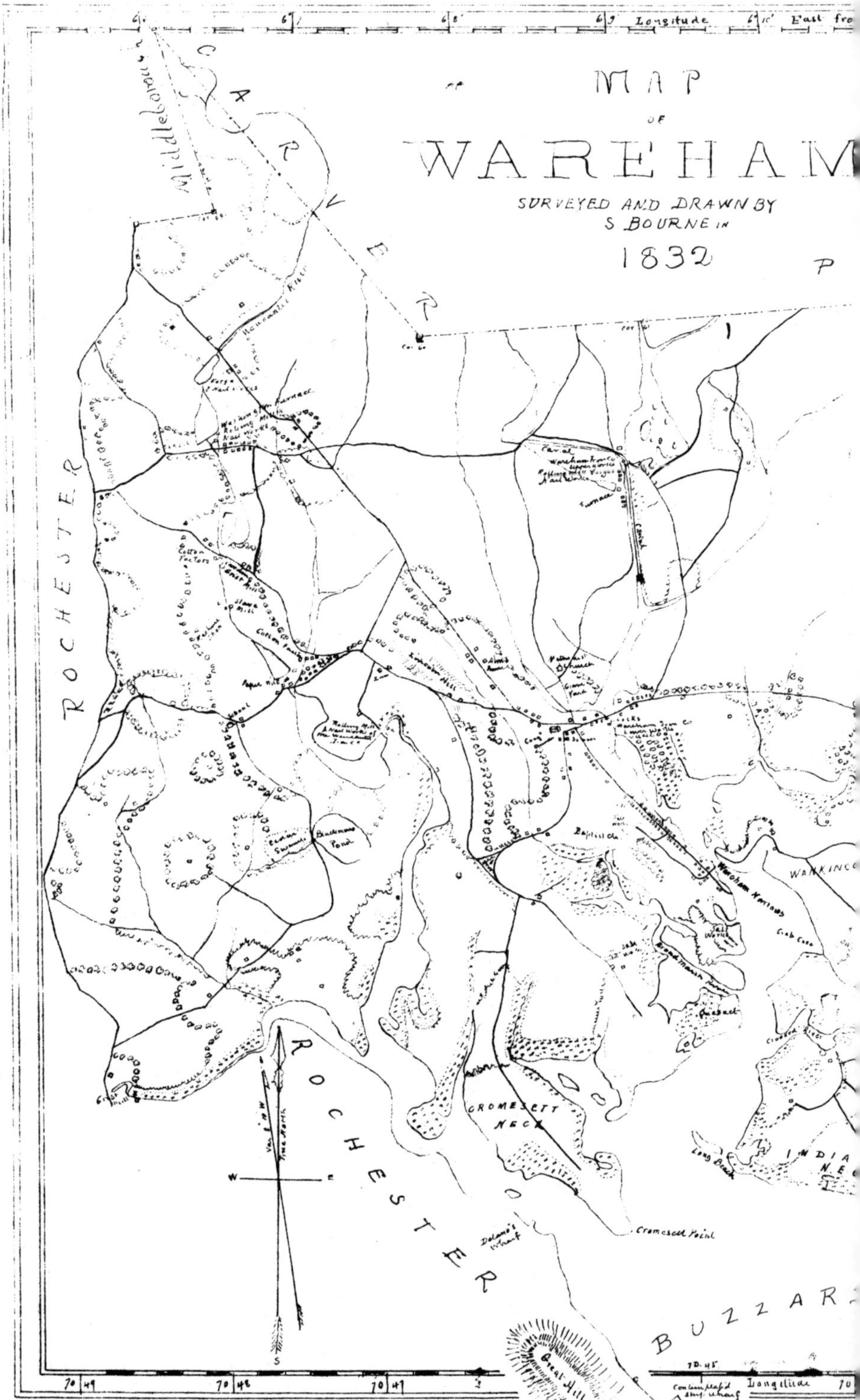
MAP
OF
WAREHAM
SURVEYED AND DRAWN BY
S BOURNE IN
1832
Longitude
CARVER
Middleborough
ROCHESTER
ROCHESTER
Blackmore Pond
Baptist Ch
Wareham Narrows
GROMESETT
NECK
Cromesett Point
Long Beach
Great Hill
Delano's Wharf

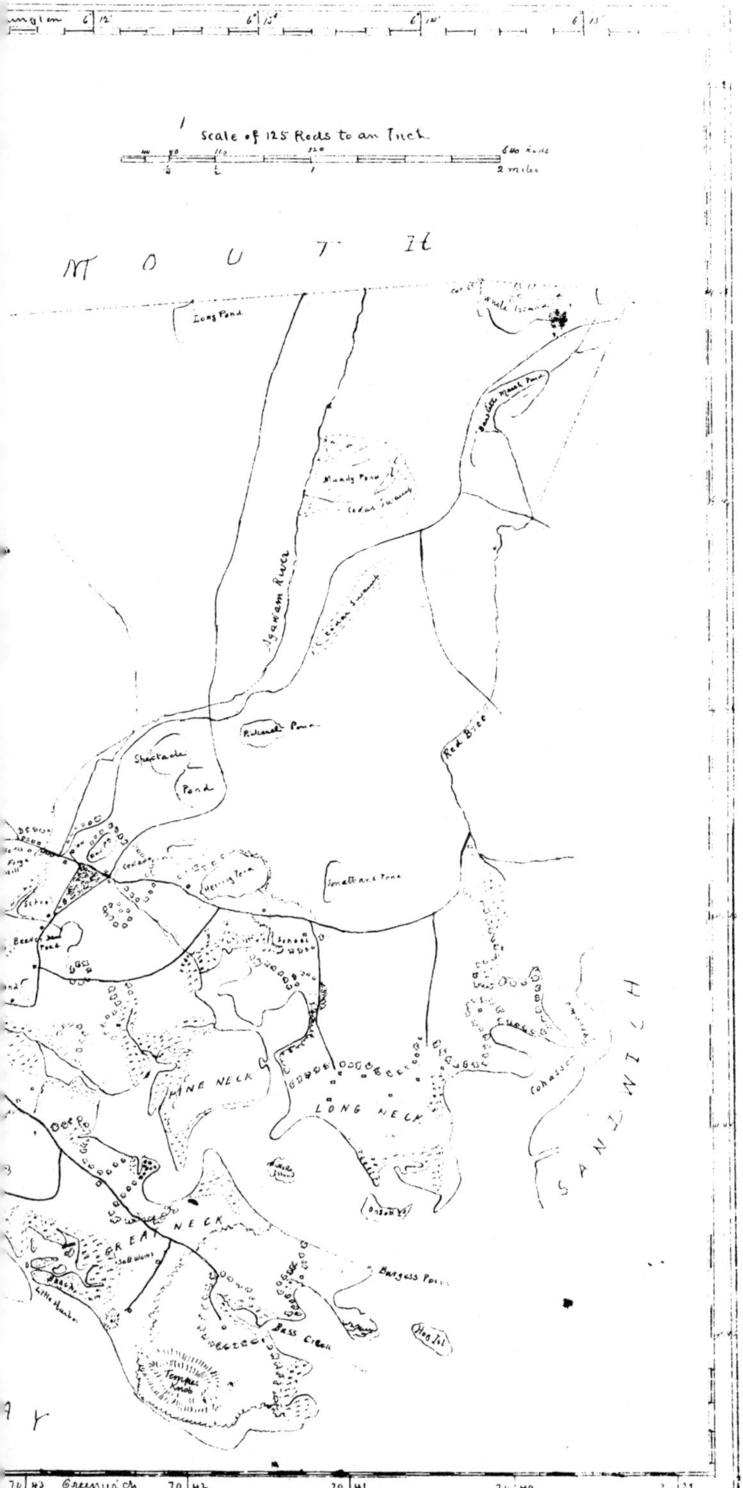
Scale of 125 Rods to an Inch
640 Rods
2 miles
M O U T H
Long Pond
Agawam River
Muddy Pond
Cedar Swamp
Pickerel Pond
Spectacle Pond
Herring Pond
Red Brook
School
L I N E N E C K
L O N G N E C K
G R E A T N E C K
Burgess Point
Bass Creek
Tempest Knob
S A N D W I C H
Greenwich

Neck, Wonkinco River, Wonkinco Neck, and other listed names, were current at the time Squire Israel Fearing and Benjamin his son lived. An earlier map, dated 1795 shows earlier layouts and names. This is the oldest map known to me. It was surveyed by Sylvanus Bourne, who invented a surveyor's mechanical measuring device resembling a wheelbarrow. This device can be seen today at the Fearing Tavern. It automatically calibrates distances as the device is wheeled down an established line from point to point, eliminating tedious chain-measuring.

Many of the old houses are located on this map. Although specifically identified owner's names are not shown, deed descriptions are very helpful in pinpointing the locations.

Salt works are shown, and churches and schools of these early times are easily located.

Main Street, known then as Water Street, is plainly visible as well as High Street, where churches and schools are shown.

An earlier map, dated 1795, shows earlier layouts and names.

A map of the 4 sea lots (preceding page).

The Four Sea Lots area was the cradle from which much of Wareham's history began. This is where the Bumpus family and the Fearing family joined to initiate (unknowingly) the center of early Wareham history. The meeting house, iron works, fulling mill, saw mill, grist mill and Fearing Tavern, Wareham's first Post Office and seat of town government, were all built on the four Sea Lots laid out to John and Isaac Bumpus, who built the dam across Wonkinco river to power their mills.

Land was purchased from John Bumpus for the first meeting house located on the center park area(as it is known today). The first meeting house was a small crude building, soon removed amidst much controversy and some violence to the area the present Congregational Church now occupies; several fires consumed the early Congregational churches until finally the present stone edifice was erected.

Maps of Parts of Old Wareham

On many of the old maps included in this history, "progress", as it is sometimes called, has led to many geographical changes. These can be noted by progressive map changes.

Brooks and streams were altered in their courses and widths; certain sections were laboriously dug out by hand to store water to run a saw mill, grist mill, cotton mill, or whatever kind was needed. These ponds had local names to identify them. Other streams were blocked by huge dams backing water upstream for

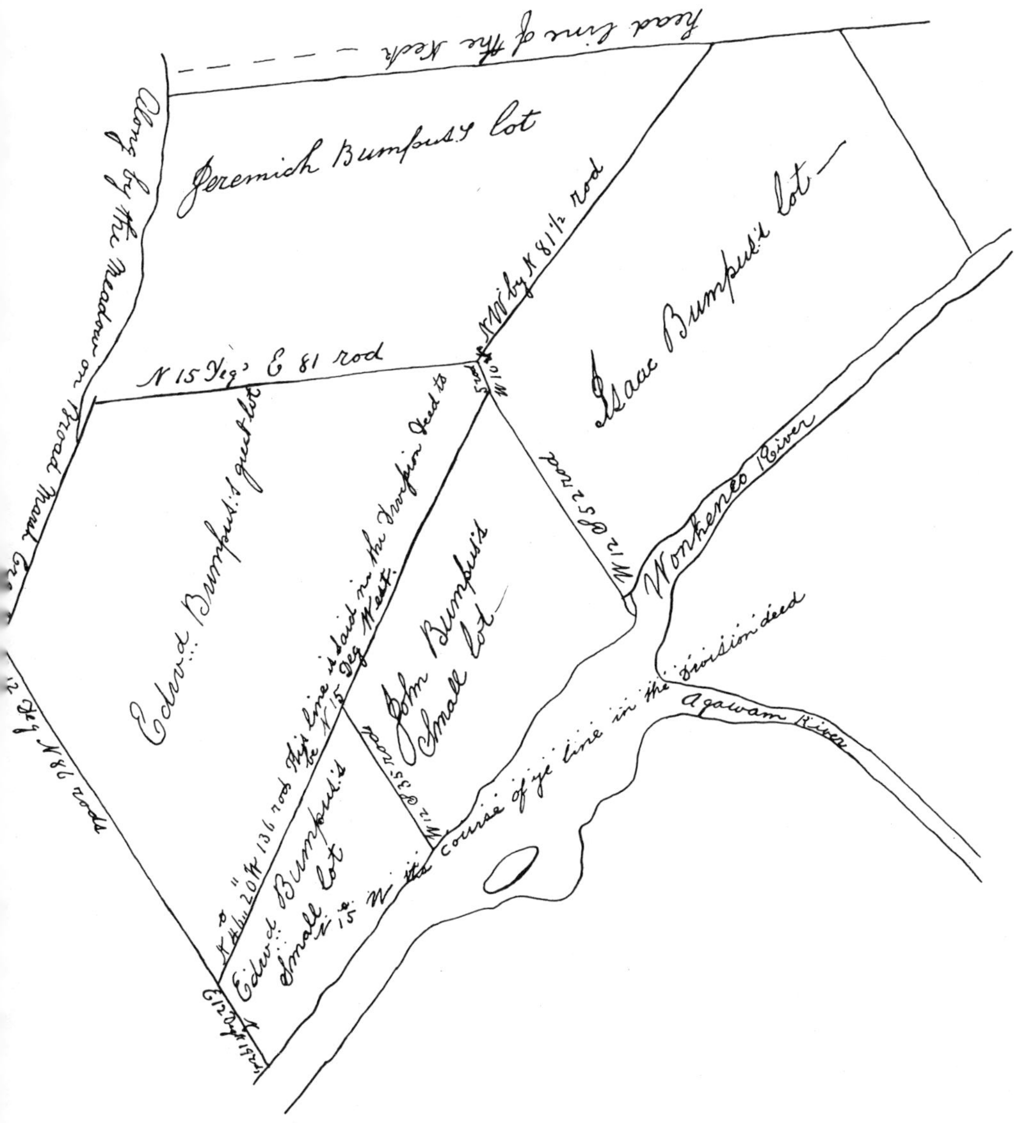

Four Great Sea Lots

miles to form a huge source of continuous water power.

On the 1749 map showing the four sea lots laid out in 1685 to John and Isaac Bump, an arc drawn on the east side of the fourth sea lot represents a small run of water, being the lower part of Bump Brook.

The railroad coming down from Boston finally entered Wareham, causing considerable dislocation as it followed the Wonkinco river bordering the four sea lots and going down to the "Narrows below Tims Island to cross the Narrows" in order to proceed to Sandwich. In its journey through Wareham, Parson Thacher's house was moved (after removing the main chimney) to a point across Main Street on the north side where it sat for several years, when it was again moved to a new foundation on Cranberry Highway, or Route 28.

This was only one upheaval; the second was the elimination of the lower portion of Bump Brook. The brook bed was extended eastward where it enters the Wankinco pond above the dam in order to utilize its moderate water flow. This cut its natural flow to the lower side of the dam and eliminated a probably costly path under the railroad bed. Careful analysis of the flow from Bumpus pond (now a cranberry bog) to the west side of Wankinco Pond shows a hand dug ditch to accomplish this purpose.

There is now a town storm drain crossing the remaining part of the Third Sea Lot to which the Historical Society has title and on which sits the old Methodist Meeting House. This drain is on the north side of the meeting house and the "little run of water" course has ceased to exist at its natural lower end.

The old town office building sits on part of the Fourth Sea Lot. The park and Congregational church are situated on the Third Sea Lot.

These lots at one time were mostly meadow, and so labeled in the history of Isaac Bump's ownership.

1827 Indenture and Agreement to

Isaac Pratt, Jared Pratt and Charles Wilbur,
Gentlemen, for a nail factory.

> This indenture made and entered into this third day of November in the year of our Lord, One thousand eight hundred and twenty seven, by and between Benjamin Fearing of Wareham in the County of Plymouth, Esqr. of the first part, and Isaac Pratt Esqr, Jared and Charles Wilbur, gentlemen, of said Wareham of the second part, witnesseth that the said Fearing on his part, *in considera-*

tion of the sum of Three hundred and seventy dollars and ninety cents to be hereinafter annually paid on or before third day of each and every succeeding month of November by the parties of the second part, so long as they their heirs and assigns shall occupy and improve the property, rights and privileges hereinafter described, hath granted, leased, and let to them the said Isaac, Jared and Charles their heirs and assigns the use occupation and improvement of the mill dam on and across Wankinquin River near my dwelling house in said Wareham together with the use occupation and improvement of two third parts of the water of said river, and the pond raised and made by said dam, and the land flowed by said pond or that may be flowed by said pond with the dam at its present height, with the right to use occupy and improve said dam river and pond for the purpose of operating and driving any works except mills or mill for grinding corn or grain, and the right of making and maintaining said dam and flowing without claim of damage as nigh and far as stipulated in a grant from me to said Isaac, Jared, Charles and Daniel Crane dated the thirteenth day of December A.D. One thousand eight hundred and twenty-one, and also the following described parcels of land situate in said Wareham, bounded by and described as follows, to wit, I. a lot of land lying on the Southerly side of said road leading from my dwelling house to Agawam Furnace in said Wareham, beginning at a stake and stones standing on the Southerly side of said road fifteen links from the Northwest corner of the front yard fence of said Isaac Pratt thence South twenty-two and three-fourth degrees East in the line of said fence and buildings eight rods and two links to the Southwest corner of the woodhouse, thence North sixty and one-half degrees east twenty rods to stake and stones, thence north fifteen degrees east to a large rock under one side of the Pecking Machine thence by the river down stream to the land Bartlett Murdock Esqr sold to said Isaac, Jared, Charles and Daniel Crane, thence East North East twenty and an half rods to a stake and stones, thence North twenty-eight degrees West to the road, thence Westerly by the road and dam to the first bound, with the privilege of cutting a canal twelve feet wide below the dam to the river to carry off the water. II. Also a lot of land bounded Southerly by the road aforesaid, Northerly and Easterly by land of Nathaniel Crocker, and Northerly and Westerly by the pond aforesaid, there are now standing on said lot a blacksmith shop and shed. III. Also a lot of land on which stand a blacksmith shop and coopers shop, bounded northerly and easterly by said pond, southerly by said road and westerly by the fence as it now stands on the east side of land occupied by

Isaac Fobes and the said Benjamin, hereby relinquish all claim to rent, past and future named in the leases from me to said Charles, Jared, Isaac and Daniel dated December 13th, 1821, and August 23 A. D. 1824, *and the said Isaac, Jared and Charles on their part undertake and covenant with the said Benjamin His heirs and assigns in consideration of the premises to pay a rent for the premises the sum of three hundred and seventy dollars and ninety cents to be paid each and every year as long as they shall occupy and improve the premises,* to be paid on or before the third day of November to him the said Benjamin, his heirs and assigns, and if the said Isaac, Jared, and Charles their heirs and assigns shall suffer said rent to be in arrear sixty days after the same shall become due and sixty days after demand shall be made on them for the payment thereof by the said Benjamin his heirs and assigns by him or them a form in power by him or them duly authorized to receive and discharge the same, then all the rights privileges and inmenitues [?] granted by the said Benjamin herein to become forfeit and the same fully and absolutely to revert and become vested in him the said Benjamin his heirs and assigns, said Benjamin his heirs and assigns to receive said rent in Wareham. And the said Isaac, Jared and Charles further covenant and agree with said Benjamin his heirs and assigns that they will not keep on the premises or in the neighborhood a house of public entertainment, nor on the dam aforesaid any mill for grinding corn or grain, there are standing on the aforegranted premises, a number of buildings the property of said Isaac, Jared, and Charles. In testimony of all which we hereto interchangably set our hands and seals the day and year first above written . . .

CHAPTER VII

The Mills

There were several kinds of mills in Wareham, all operated by water-power. The first mills were saw mills.

The earliest records of up-and-down saw mills are in York, Maine in 1623; Portland, New Hampshire, 1631; Scituate, Mass. Bay Colony, 1680; and Silver Mine (location unknown) in 1688.

John and Jacob Bump had a saw mill on the Weweantic River working about 1670 and later had a saw mill on the Wankinco River at the dam site. These saw mills operated a saw with 2-inch teeth that went up and down in a recessed frame, with greased blocks attached to the hardwood frame allowing a minimum of friction as they moved in the recesses of the fender posts. A crank and pitman rod attached to the saw frame was powered by a water wheel mechanism. Although very inefficient for the power obtained, it was nevertheless faster and less exhausting than sawing by hand.

The circular saw was conceived by a Quaker woman in 1810, Sister Tabitha Babbit of the Harvard Quaker Community, while watching her spinning wheel. The circular saw revolutionized the saw mill by speeding up the sawing process and leading to other inventions such as the planer and wood-trimming devices. The speed of the up and down saw was 120 strokes per minute.

The speed of the circular saw was many times the speed of the up and down saw. In addition, the circular saw developed a centrifugal force which added more power to the cutting process.

The second most important mill was the grist mill, of which Isaac Bumpus was known to have at least two and probably three. One mill was located on Bump Brook, now known as Rose Brook. One grist mill was on Wankinco River, and possibly a third on Weweantic. The remains of the Rose Brook mill were evident as late as seventy-five years ago (remembered by Daisy Lovell, author of *Glimpses of Early Wareham*).

Weaving was originally performed at home on a loom of early construction, but as soon as cotton mills were furnished with mechanical spindles, home weaving began to rapidly disappear. This came to pass about 1812, when Jonathan Reed erected the first cotton factory on Wankinco River dam, the same factory set afire by the British when they entered the Wareham River with six small boats from the Bark *Nimrod*, a British gun-ship, during the War of 1812. The fire in the factory was put out after the British left.

In 1815 another cotton mill was erected by Curtis Tobey of Sandwich on the Weweantic River near Fearing Hill. In 1826 this mill, built of stone, was leased to Bartlett Murdock of Carver and Joshua Tobey for $350.00 per year.

In 1823 another cotton mill was built by Benjamin Lincoln of Norton, and Curtis Tobey, on the Weweantic River near Bump Bridge. It started with 420 spindles.

In 1824 a paper mill was started by Pardon Tabor on the Weweantic near County Bridge.

A stave mill owned by Seth Tobey was operated on a small stream spilling into the Weweantic River. Basically this was a saw mill but specialized in making staves for kegs and barrels.

Iron mills were operating on the Agawam, the Weweantic and Wankinco Rivers.

Several nail factories were operating at the same time. Furnaces and nail factories proliferated. The Town of Wareham became very busy and prosperous, but like all good things it did not last. Along toward 1834 the iron industry began to decline and rapidly moved to extinction before the twentieth century, when cranberry bogs were built in bog areas where iron ore had formerly been dug.

Grist mills had long been in operation and were in 1830, but began to have considerable competition from other more efficient mills. The Old Colony Railroad creeping down from Boston year by year soon closed nearly every type of mill then operating. In 1848 nearly all the local mills had ceased operation, and the work force began to leave Wareham.

It is difficult to understand why the Fearings never invested in the manufacturing process, save for Israel, who bought the holdings of Isaac Bump, who had in turn owned several kinds of mills. Benjamin, who inherited all the Bump property, did not himself operate any of these mills but wisely leased them to others.

The only remaining grist mill in this area is at the Hiller Farm in Rochester, Mass., on Mary's Pond Road. The mill is located on

The Old Grist Mill, South Wareham, Mass.

Standard Horse Shoe Works, South Wareham, Mass.

the east branch of the Sippican River, fed by Leonards Pond, a pond created by a dam erected for water power development.

Over the years (before 1841) the water wheels were repaired and maintained until the time of the invention of the water-turbine. This device delivers greater power by reason of a large column of water directed to blades on a shaft connected to the machinery. This turbine could be operated from inside the building, and occupied a very small space compared to the enormous one heretofore needed by the water wheel. Its placement on the outside, where freezing water had constantly injured the old wheel, was also remedied.

The main floor of the present mill is much as it always was, with the huge grinding stones enclosed with only an entrance for the grain and an exit for the ground meal, the latter channeled to fill the boxes and bags on a lower landing. The whole grain was lifted to a second floor from which it followed connecting wooden pipes from the hopper to the grinding wheels below, to assure the most economical procedure and prevent loss of grain from spills.

Belts and gears attached to the turbine driving shaft moved the grinding stones and other machinery, including a waterpump to deliver household water to a storage tank in the farmhouse.

Another shaft and pulley later drove an electric dynamo for power and lights on the farm, and operated tools for repairing and sharpening farm instruments.

The smell of grain is still in the air and old boxes and barrels — never used — are still in the loft, together with wooden pipes and other necessities for operating the grist mill.

Mr. Howard Hiller, owner of this ancient grist mill, proudly showed the author around and through the wonderful old mill, a haunting stamping ground of characters from the 18th century. The farmers of Rochester and Wareham used this mill. In Mr. Hiller's memory he recalls several who came to him to grind their grain into meal. Mr. Howard Hiller, himself, was the last to operate this mill, built by Peter Blackmore as recorded by Samuel Prince.

The Leonards, who owned this property at one time, came from Taunton (as did several others) to operate mills, both iron and cotton, in Wareham.

This old Grist Mill was on the Weweantic River in the late 1800's. Only a raceway on the bank of the river and two millstones lying in the stream remain as mementoes of this mill today.

The buildings of the Standard Horse Shoe Works have been torn down, and only the old foundations remain. The land was purchased recently by the Conservation Commission of the town of Wareham for recreational purposes. The site is on the Weweantic River, near the end of Mill St.

Because the Fearings were involved in the production of salt at a salt works, it is fitting to explain what a salt works is and how it operates.

These salt works were located at Nobska Point, Barney's Point, Little Harbor, Griffin's Field, Tom's Narrows, Henry's Creek, Old

Pan, East and West sides of Muddy Cove, and other places along the shores of Wareham.

One method of obtaining salt from sea water was to boil the sea water in large iron kettles set in stone or brick on a more or less permanent basis. Each location would make about 30 - 40 bushels of salt a week. As the salt works operated six months a year the total of all the works would reach 8140 bushels a year at one dollar per bushel.

About 1806-1807 a faster and cheaper method was invented: vat evaporation in the sun. Large flat surfaces of wooden construction were built to hold the sea water, then exposed to the evaporating rays of the sun. This process began with Asa Swift and Nathan Doty who built 1500 feet of open vats (a foot of salt works was 10 feet square). Hallet Swift built 1000 feet at Pigs Point. Covers were later built to keep out the rain.

At various places from 1821-26, Aborn Gibbs built 1000 feet, Stephen Swift 500 feet, Peter Smith 1000 feet, Ichabod Leonard 2000 feet, Benjamin Fearing and Stephen Gibbs 1000 feet. By 1911 no traces of the Salt Works remained.

In 1807 fifteen cotton mills operated in New England. By 1811 there were 87 mills with 80,000 spindles. Four years later 500,000 spindles in the country were attended by 76,000 workers.

Transportation by rail had expanded enormously, bringing products rapidly from distant points to the new population centers.

Several machine-production crashes occurred, one after the other, first in 1819, then 1829, 1837, after which a brief recovery, another crash coming in 1847 and a fifth in 1857.

These several depressions had devastating effects on the old mills in the former colonies, and Wareham was no exception. These crashes followed the War of 1812.

Wareham was essentially an iron town, although other small industries such as saw mills, stave mills, grist mills and paper mills existed.

The Civil War ended in 1865 and a panic followed shortly afterward. The mills had begun to close in 1848 as cheaper transportation, low cost power and other conditions pressed the mills to close. But as the old iron mills closed a new industry was born and grew to prosperity — the cranberry industry, with a second built upon the sparkling silver beaches and their warm water and plentiful fishing, bringing summer visitors from all over the nation to give the economy a new incentive to growth.

Pennsylvania had become a highly industrial state during the Civil War, growing greater each year thereafter. The industrial base had been laid long before, when coal was discovered along with iron in quantities sufficient to make Pennsylvania the leading industrial state, out-pricing and out-producing Massachusetts. It forced the Massachusetts Colony iron works out of business in 1848, and left only the Washington Steel Co. to produce on a limited scale, finally closing after a period of decline. In effect, the older works could not compete with new and more modern methods from other states.

The iron workers gradually moved to other more prosperous areas and new people came to populate Wareham. Cape Verdeans and Italians came to build the railroads and work in the new cranberry bogs. The Cape Cod Canal, built from 1909-14, attracted many of the new immigrants and changed the population mix for all time, with the result that Wareham now has a diversified population, but few industries.

With the loss of the iron mills, a great source of wealth and employment, mill owners began to look for other investments. The cotton mills, grist mills, saw mills and fulling mills closed down as more modern mills came into competition with them.

Even agriculture in Massachusetts received a blow as the western prairies were plowed for the first time, producing abundance in the deep virgin soil of this general level area covering several states.

CHAPTER VIII

Furniture Period 1690 - 1900
Early New England
Imported and Homemade

Not all of the early settlers were impoverished; some were rather affluent. They came to the new land not because of religious preferences but to avoid taxation. From the inventories of small businesses, we learn that few pieces of furniture and many times none were listed in the inventory of the owner at his death.

The best of the furniture was either imported from England and other countries or purchased from local artisans. Most early homes contained a very few pieces of furniture, mostly beds, tables and benches, especially in Pilgrim homes.

Frame houses were built as early as 1628, so it is easy to believe that rough furniture as well as houses were made from sawed lumber from this date.

One of the most ingenious pieces of furniture was a "settee". It consisted of a hinged seat covering a box-like storage bin and a very high back. When this piece of furniture was moved up close to the fireplace, it was quite comfortable in that the high back reflected the heat onto the back of the sitter, shielding him/her from the cold blasts of air from behind. They were quite popular, as a large number of them exist today.

Benches at the tables were common for eating arrangements. Several varieties of stools, both low and high, and lacking a supporting back rest, came into use.

The original crude seats or stools were gradually improved as time went on. A simple upright board was fixed to the broad stool to make the first chair. Then a variety of turned and flat upright back braces were fixed at the top by a cross piece or a curved top, depending on the skill of the husbandman.

Over the years large storage chests, "Captain's chests" of various sizes and trunks were constructed to store the season's clothes until needed. Casks, barrels and boxes called "firkins" were added to the storage facilities and small carrying conveniences began to grace the small homes.

Wooden ware, easy to make and replace took the place of the glassware left behind in England, and was to be replaced by hollow ware constructed and formed by iron workers of a later period.

The more affluent settlers accumulated pewter plates, cups, pots and serving dishes. Coin silver teaspoons, linen, looking glasses, window cushions and many other comforts were acquired from foreign visiting vessels, traders and merchants.

Cat-tails and turkey feathers were used to stuff the mattresses which were laid over rope supports woven from side to side of the bed frame and tightened by winding a lever at intervals. To have a complete feather bed mattress was considered heavenly.

Cupboards and corner cabinets became necessary as people accumulated more luxuries over the years.

Small root cellars to keep apples, potatoes, turnips and cabbages were dug under the house with an inside entrance to keep the severe cold from penetrating the cellar.

Books were very rare, except in the houses of such people as the minister. Rev. John Norton had, in his inventory made after his death, "729 books, plate, a case of drawers containing English and Spanish coins, a clock and case in the parlor."

One of the craftsmen of the times had listed in his shop for sale after his death, 4 boxes, 3 chests, 2 bedsteads, chest with drawers, boards, planks, timber and joiners tools with a value of 20 (pounds).

About this time, 1633, joiners and carpenters began to come to the Colonies who were not identured but were given surities by wealth men who employed them or set them up in business to produce furniture or build houses.

Rush bottomed and caned chairs begin to appear in 1646, made by one Christopher Stanley. Leather bottomed chairs appeared earlier. Rocking chairs appeared about the same time as cushioned chairs (1646-9). Most of these latest additions were found only in the homes of the wealthy, the poorer colonists continuing to use the rough furniture with some simple improvements.

The principle woods used to make furniture were oak, ash, elm, walnut, maple and pine; red cedar and swamp cedars were also frequently used.

Skilled craftsmen were among the immigrants — not merely

1820 Kitchen, Fearing Tavern
Showing Settee and Long Table

carpenters and housewrights, but turners, joiners, cabinet makers and carvers.

In 1630 the pay for a skilled laborer was sixteen pence per day — about $2.50.

To keep the indigent and poor out of the colony somebody had to go bail for the newcomer even though he be a skilled craftsman.

Chairs were not common — most people used the settee or bench called a form. Most chairs were children's high chairs little changed up to the present time.

About 1650 a number of types of chairs began to be produced — some covered with leather, some caned, and some cloth covered. The couch began to appear in various treatments and lengths.

Round tables began to appear around 1650 and were made of oak, pine, maple and other woods, with both plain and carved legs.

After 1650 the homes of the wealthy began to fill up with chests of drawers, massive oak chests, chests on chests, cupboards and combinations. Many of these pieces were ornately decorated with carvings and recessed panels surrounded by elaborate mouldings. Most were imported from France and England.

Desks of numerous designs with and without legs, some with magnificent carvings and hinged sloping tops were made.

Book-cases began to appear in the homes of the wealthy around 1650.

Closets, corner cabinets and china closets became popular as more and more dishes and pewter utensils became common.

Trunks and footwarmers became more popular.

By 1751 clocks were quite numerous, from 1650 on having shown an increase.

In 1734 clock makers were active — Aaron Miller of Elizabethtown advertised in 1747. In 1749 Moses Clement is listed among Broadway, New York City advertisers.

Typical contents of dwellings previous to 1650 might be listed as follows:

Period Furniture, Fearing Tavern,
East Parlor

"Armes and tooles"
"Aparell" (Apparel)
"Canvas to make sheets with Bed & Bolster to Fill"
"Rugge & Blankets"
"Household Stuffe"
"One great iron pot"
"Large and small kettle, skellets"
"Frying Pannes"
"Gridiron, Spit, Platters, Dishes, Spoons, Knives"

The dwellings, indeed, were almost bare.

In 1637 Adam Lindsey of York "died possessed of only one block bed and covering valued (at) 80 lbs of tobacco." Anthony

Panton's estate contained bed-board, one brush, and one chest. Edward Bateman, "carpenter of St. Marie's," possessed a "boat tool, two bands, a tinderbox, a brush, a rope and old doublet, a bearskin and a chest." These were valued at 345 lbs. tobacco.

These instances are typical of indentured servants freed of their bondage, whose possessions indicated the bareness of life and the extreme poverty of their kind.

Even the more affluent possessed little more than "Justiman Snow late of St. Mary's, planter." He listed "knives, nails, smoothing irons, tools, spades, pins, line thread, ribbon, stuff 'friz' canvas, button hooks and eyes, shot, nets and lines, boats, weapons, trunk, chests, wearing apparel in all stages of decay."

"Pipes, beads, household linen, provisions, cooking utensils and livestock, 2 looking glasses, 3 dozen of trenchers, wooden platters, flock bed, feather bed, one bed standing in the parlor, the bedde and the appurtenances in the little parlor, a parcell of books, a parcel of old household stuff, 3 kettles, a chest and chayer with other household stuff, the bedding chest and tubbs in the chamber."

Some rugs were made of leather.

Oak furniture was imported from England circa 1630 and after. Mahogany was used very little until 1730, when this wood began to be used in tables, dressing boxes and other occasional pieces.

Molded panels, hand-made, began to appear in the better homes, especially after the Revolutionary War. In the upper chamber of the Fearing Tavern two excellent examples of this painstaking and beautifully executed work of large panels, fluted columns, recessed closets and window seats can be seen today as a living memorial to the craftsmanship of Colonial Times.

Wide floor boards were common in the early days, up to twenty-three inches wide. They were laid square edge to square edge, but as time and shrinking took place gaps as wide as a half inch opened and were dirt and dust catchers.

Many furniture pieces have been earlier listed. But the most popular piece of furniture was the table, about six feet long, on which thrice daily large amounts of food were set out by the Colonial mother for her hungry husband and their large brood of children, whose appetites consumed enormous amounts of food required for the performance of their daily chores.

A sale by auction on May 12, 1823, of the contents of one of the Bumpus family households points up the nature of the stark, crude furnishings of the period.

1 foot wheel — to Robert Besse	$ 1.10
An old spinning wheel — to Alicia Barrows	12.50
A doubled geered wheel — to Selte Sturtevant	1.10
One Bed — to Samuel Ellis	2.00
2 Bed Blankets — to Col. B. Murdock	2.65
2 privy shacks — to P. F. Briggs	2.00
1 pr. pillow cases — to Col. B. Murdock	.40
1 cover lid — to — —	2.00
1 Old Bed quilt — to — —	.25
1 Bed Bolster & Pillow — to Elija Bumpus	10.00
A lot of gown and lock — to Samuel Ellis	3.65
5 pr. stockings and pr. mitts — to P. L. Briggs	.85
4 pr. Cards — to Alicia Barrows	.21
Lot of Pewter — to Enos Howard	.70
Sundries — to Wm. Mackie	.85
1 Bible — to Curtis Tobey	.35
2 Chests — to David Swift	1.70
One of Ironware — to Capt. Jeremiah Bumpus	1.60
One bedstead and Rope — to Capt. David Nye	.28
One table — to Wm. Mackie	.20
2 lb. Sheep Wood — to Samuel Ellis	.80
1 lot of Churn & Lamekens ? — to Robert Besse	.90

Taverns or ordinaries of the 17th century were far from luxurious in their furnishings.

When one known proprietor died his estate consisted of the Inn and the contents as follows:

"In the kitchen three formes* dresser and shelves."

"In the larder, one square table, banister, dresser and shelves around."

"In the hall three small rooms with tables and benches in them, one table about 6 feet long in the hall and one bench."

"In the low parlor one bedstead, one table and two benches, two formes, one small frame of a form and shelves, one chest with shelves."

"In the room under the closet one child's bedstead."

"In the chamber called London one bedstead, two benches."

"In the chamber over London, one bedstead, one Crosse

* *High backed settees.*

table, one form, one bench."
"In the closet next the exchange, shelves."
"In the Barr by the hall, three shelves, the frame of a low stoole."
"In the upper parlor, one bedstead, two chairs, one table, one form bench and shelves."
"In the nursery, one Crosse table with shelves."
"In the court chamber, one long table, three formes, one livery cupboard and benches."
"In the Starr chamber, one long table, one bedstead, one livery cupboard, one chair, three formes with benches."
"In the garret over the court chamber, one bedstead, one table, two formes."
"In the four garret chambers, over the Starr chambers, three bedsteads, four tables with benches."
"In the Brew House, one cop, two falts one under the back, one upper back, one kneading trough, one dresser, one brake."
"In the stable, one rake and manger."
"In the yard one pump, pipes to convey to water to the

brew house, tyne hog styes, one house of office."

"The sign of the King's Arms and signe posts."

This long list of seemingly numerous items, on further consideration, will indicate a very sparse furnishings of each room, giving a barren effect. Quilts, sheets, pollow cases, comforters, books, even the Bible are not recorded.

CHAPTER IX

Wool, Linen, Food, Clothing and Work Habits

All flax and wool yarns were spun by the women and girls. They made them into sheets from which clothing and bedding were stitched, all by hand.

Some of the women became quite expert in making patterns in the cloth from varied colored threads on treadle looms. The home looms soon became obsolete as improved cotton and weaving mills became operational.

The flax was raised in a special plot from which the men pulled, rather than cut, the stalks to save every bit of the flax. When the stalks were well dried they were threshed for the seed. The flax was then spread out until all the pith and woody parts became brittle with decay.

The timing had to be precise so that the fiber did not become rotted. The fiber was the part saved. When the right degree of the decaying process was reached the flax was then tied into bundles. The next process was to "break" it on a clear cold day, the best time for this operation. A rough cogwheel-like machine was made for this breaking, which made a loud noise as heavy strokes completed the process.

The flax was then held in small portions in the left hand over the edge of the board and beaten with a thin stick called a "swinging knife." It was beaten on one side, then turned over and beaten on the other side until it became clear of all the bits of stalk. This process was called "Swingling".

Now came the woman's turn to finish the preparation. On a long plank a brush made of iron or steel spike called a hatchel was attached. The flux was combed evenly and smoothly over first a coarse hatchel, then a finer hatchel. When it was done it was twisted into what resembles a hank of very long dark hair.

The long hanks were wound on a distaff and made ready to be spun on "a little wheel".

It was wound on a spool directly as it was spun, hanks being set in a half oval structure which circled around the spool. The flax was pulled off this distaff a little at a time, wet occasionally from a dish set for that purpose.

After the spinning came time for weaving this fine thread into linen on the treadle loom. A curing process was applied to the finished cloth by spreading it out on the grass after boiling the cloth in a mild lye solution made of ashes and water, and keeping it constantly wet until the process was finished.

The clothes of the early colonial settlers were made of rough cloth similar to canvas with a coarse weave called "fustian." Those who were Quakers, for instance, eschewed all gaudy decorative clothing.

Heavy cloaks and hoods were worn by women pioneers. They brought large amounts of clothing with them on the voyage from England. This clothing rapidly wore out from the labors and rough living endured by these early settlers. Consequently, clothes made from homespun cloth took the place of the original clothing.

Clothing was initially supplied in great measure by the company of London merchants who supplied the ships, food and early instruments for setting up the colony. More than likely the clothing was of a uniform pattern made for durability and not for style. The Pilgrims' clothes were much different than those of the Puritans.

The men wore doublets, leather jerkins, breeches, and great three-quarter length cloaks. Linen was worn around the wrists and neck and the sleeves were fairly close to the arms. For head gear the Pilgrims wore wide-brimmed hats with a high crown.

The shoes were made of leather, heavy and strong, and came up to ankle height. Boots of various heights were worn by some Pilgrims. Heavy woolen stockings covered the feet and legs to a height above the knee.

In addition, the men wore swords, powder flasks and guns. Heavy, serviceable gloves were worn, not carried. Canes were used by older men.

There were four classes of people, generally identified as:

1. The educated and merchants, who were the clergy and business people and controlled the colony.

2. The merchants and merchant adventurers, who in time dominated the trade routes, warehouses and systems of distribution.

3. A layer of society made up of Yeoman tenant farmers, freemen, servants and sailors.

4. A group of African slaves whose dress was anything discarded by any one of the other groups, fitting for their stations in life.

As we ascend the scale of estate, clothing became more elaborate and was made of finer materials, especially that of the rich merchant class, who had access to all that Europe could offer. Wearing these gaudy and expensive clothes was one way of showing their superiority over others.

In the days of Benjamin Fearing's Tavern, food was plentiful and varied from season to season. Wild game abounded and domestic animals were more than adequate, as overgrazing on public lands was often a grievance among local farmers. With few exceptions the needs of the family were grown on the farm land. Wool to weave cloth for clothing came from the sheep, and flax for linen was grown in a special garden; herbs for health potions and flavorings was carefully tended. Apple orchards were pridefully grown for variety of good eating, carefully selected, and packed in barrels in the keeping-cellars for winter.

Malt was dried or salted to preserve it or, for the most part, to order at any time of the year.

Land was sown to oats, wheat, barley, corn and rye: threshed, then carried to the grist mill to be converted into grain after saving enough seed for the next planting season.

Ducks and geese were table favorites the year round. Wild turkeys were often eaten also.

The innkeepers always had a good variety of game, fish and fowl for their guests.

Sugar and molasses were imported from the West Indies in exchange for tar and turpentine "milked from the trees" solely by permission of the early proprietors. Later land owners could do as they pleased, especially in the period after the Revolutionary War.

The herring "running up and down the Wankinco River" were a source of food when they returned to spawn on the upper parts of the river, going up the fish ladders provided to get around the waterfall of the dam.

Later the potato was popular as a staple food item. It came to England in 1596 and was later introduced into New England. This tuber was known in England under several different names. Its easy propogation, even in short growing seasons such as New

England, made it a popular food item. It could be easily kept in edible condition through the cold northern winters buried in the dirt floors of the keeping-cellars, and in barrels and boxes, as with apples.

Work Habits

The whole family of farmers' sons, daughters, and indentured servants took up their daily task before sunrise, pausing only for meals, and resuming their labor until the candles were put out for early bedtime.

The women did the house work. They also took care of the domestic animals, scoured the pots, pans, warming pans and pewter dishes. They spun flax and woolen yarns and wove them into cloth from which clothing and bedding were made by their own hands.

If there was a fulling mill near by, the rough woolen cloth was finished and cleaned by the mill.

The women also prepared the family's meals by cooking and baking in the huge fireplace, some having an oven on each side of the hearth.

Washing clothes and caring for children was a continuous operation, since it was common to have a large number of children (many died with their mothers in childbirth or shortly thereafter).

It is noteworthy that some men married three and four times because of the early death of their wives from disease or childbirth or both. Few women outlived their husbands at this early stage of colonial society. It was a rugged time of fearful toil, for men and women both.

The man worked in the woods cutting firewood, "milking the pines for tar and turpentine", and clearing areas for more planting or for building houses. They toiled as blacksmiths, boat builders, sailors and fishermen, and a certain amount of them were impressed into "His Majesty's Service" as soldiers or sailors. The longer the daylight the longer the work day.

Children shared the work in and outside the home. The long daylight hours were part of their life — everybody worked except the few fortunate rich.

Occasionally one of the home-staying daughters became so skilled in needlework that her services were sought after by neighbors less skilled to employ her skills in their behalf. Moreover the rich wives of the plantation would hire her to make articles for their homes. The pay for a day's work was her "diet", 16 cents in winter and twenty cents in the summer because of the longer daylight hours.

Children were taught early to work and acquired contentment with their labors. As they matured into men and women some never crossed the boundaries of the town, while others moved ambitiously to other areas where they would become the founders of prosperous families in the commercial life of colonial towns.

By Courtesy of Frank E. Davis Fish Co.

Heaving the Seine

IN HOC FERRO VINCE
Fearing

CHAPTER X

The Fearing Genealogy

The Fearing family has a long and honorable line of ancestors to John Alden, the Mayflower passenger, and even further to a woman named Annis Fear (Fearing, Fear, Feerin, etc.), who married a John Hawke in 1650. A William Fearing of Stondon Massey England whose will was proved in 1596, and others give considerable authenticity to the early Fearings and Hawks. The Daniel Cushing Record states "that in 1638 on board the *Diligent* of Ipswich were 133 emigrants including Mathew Hawk, his wife Margaret and servant John Fearing." Hawke came from Cambridge as did John Fearing. Hawkes died in 11 December 1684. Later the two families were related by a marriage of a Fearing woman to a Hawke man. The name Hawke remained in the Fearing family as a given name for several generations.

John Fearing arrived in Hingham in 1638 or 1635 and his grandson Israel Fearing 3 married Martha Gibbs and moved to Wareham where all their children were born (see genealogy). This Israel Fearing 3 died in 1754 and left to his son Benjamin all of the Isaac Bumpus lands purchased earlier.

Through this Benjamin a son, Benjamin, who married Salome Pope and became the father of William Sturgis Fearing who married Hannah Cushing Barrows. They had a daughter Mary Fearing who married William Baily Goodall. They had a daughter who married William David Gherky. They had three daughters:

Doris, who married Dr. Gilson Golby Engel,
Jean, who married Mr. William Killhour, and
Eleanor, who married Mr. T. Rowe Price.

John Fearing came from Cambridge, England with Mathew Hawke and wife. He became a Freeman in 1652. John Fearing died 14 May, 1665.

A genealogical survey made by some of John's descendants found, through a search of early Cambridge parish records and surrounding parishes, a record of several Fear's, Ferin's, Fering's and other spellings of the Fearing name back to 1564. A Thomas Fearing, a brother of John who "if he shall come from beyond the seas," as stated in Thomas' will as a potential beneficiary, is linked thereby to the John Fearing we know as being the first immigrant from England of that name to land on these shores in Hingham of Plymouth Colony.

A descendant of John Fearing, one William Sturges Fearing, a fifth generation descendant of John's, married a Hannah Cushing Barrow whose ancestors are recorded as early as 1500 in England.

Further claims have been supported by certified data that William Mullens, born in England, was a Mayflower passenger, a citizen of Plymouth Colony, and died in Plymouth 3 March 1621.

Doris Engel has also certified her membership in the Pennsylvania Society of the Colonial Dames of America by right of descendant from Francis Chads of Chester County, Pennsylvania

Doris Gerky Engel, an eighth generation descendant of John Fearing, has certification tracing ancestry to John Alden of the Mayflower. (Vol. II pp 114, 118 Savage Vol. 1 p. 23 Descendants of John Alden 1867 pp 1-2.)

FEARING GENEOLOGY 1638-1970

The Fearings married into the Bump family, who can trace their ancestors through the Warren line back to William the Conqueror of England born in 1027, and King of England from 1066-1087.

The Fearing family on this side of the Atlantic started with John Fearing who arrived in Hingham in 1638 and died 14 May 1665. His wife was Margaret. They had several children, all born in Hingham, among whom was Israel the second son was born September, 1644 and died at Dorchester, in June of 1693. His wife was Elizabeth Wilder. They had a son, John, and a son Israel. Israel was born 29 August 1682 and he married Martha Gibbs. He resided at Wareham where he had eight children — five sons, Israel, Benjamin, John, David and Noah, and three daughters, Ann, Elizabeth, and Martha. Of these children we are immediately concerned with Israel, Benjamin and John and more particularly with Benjamin and John who had the most contact and use of the Fearing Tavern. John inherited his father's office as Justice of the Peace and became the "Squire". His brother, Benjamin, owned and operated the Tavern and the immediate lands around the area known as the Sea Lots annexed by the Town of Wareham by state legislation, taking in the Agawam purchase as part of the incorporation in 1739.

The Hingham branch of the Fearing Family continued to grow and expand as did the Wareham Fearings until the 1848s when the Fearing heads of families began to leave Wareham. There are two reasons for this, (1) the iron mills in Wareham began to shut down, bog iron becoming scare and fuel running out. The Mesabi range of iron in Minnesota was discovered along with Pennsylvania coal and the combination was creating a competition that Wareham iron mills could no longer meet; (2) The Old Colony Railroad was building from Boston to the Cape creating competition with the coastal schooners, which finally ceased operations. The loss of low-cost shipping added to an already disastrous situation and the mills closed for good. The workers moved out to more prosperous industrial areas.

The railroad passenger service closed many inns and other marginal businesses. It caused a great upheaval and change of tempo within Wareham. Stage coaches ceased operations as rail service was extended up and down the Coast. The Fearing Tavern then became a private dwelling, although some town business was still transacted there.

William Fearing became town clerk and conducted the office from his dwelling, the first Post Office also being established there in 1814, by earliest official record. Peleg Fish covered the route from Tiverton, Rhode Island through New Bedford to Sandwich twice a week. The first payment by the Post Office Department to the Wareham Postmaster was in 1825 for the sum of $83.62.

William S. Fearing, Town Clerk and later Postmaster, was the son of Benjamin Fearing and Salome Pope. She came from Sandwich. William married Hannah Cushing Barrows, whose ancestry can be traced back to a John Barrowe of England, born about 1609.

William and Hannah Fearing had a daughter Mary who married William Bailey Goodall of Philadelphia. She was born on the 8th of October, 1849, and died February 25, 1946.

Mary and William Goodall had a daughter, Hannah Fearing Goodall, who married William David Gerky. Their three daughters, now living in 1977, are Doris Engel, Jean Kilhour, and Elinor Price. These are the three descendants who gave so generously to restore the "Fearing Tavern" to its former condition as an Inn.

There are less than 1200 individual Fearing members left representing 500 families all over the country. They are a declining entity in the United States of America today. It seems depressing to see this splendid and famous family slowly dying out as generation after generation declines in number of members.

Character Analysis of Esquire Fearing

The author is a character analyst of handwriting traits and offers this analysis of Israel Fearing the second, of the third generation. He was the Squire Fearing who was born in Hingham 29 August, 1682 and moved to Wareham. His interesting character traits prove him to be a man fit for his positions as Justice of the Peace and Squire in Wareham.

He was a hard man, very suspicious, egotistical and determined to have his own way. He was intolerant of any beliefs not conforming with the established current customs, habits and modes of living. He was very selective in his friendships and aloof from the lower class of workers.

He was a clear thinker, rapid in his decisions and quick to see a business opportunity to financially improve his circumstances.

He had an appreciative eye for the ladies and was gallant when in their presence.

He loved to eat the best of food and would reject any food not cooked and prepared to his taste.

He was a sensuous man who loved his woodlands, meadows, streams and rivers. He loved to walk in the woods and was often called upon to survey the town border lines, insuring the location of "meets and bounds."

Squire Fearing was probably of above average heights but quite a heavy man befitting his station in life as a country gentleman with great responsibility.

He was a good horseman, as witness his long journey to Boston to get the necessary legislation to incorporate the town in 1739, traveling a path over streams and brooks, through winding and uncertain paths overgrown with bushes and trees, sometimes guided more by his horse than himself, a journey which took him over a day to complete.

A man of courage and integrity who faithfully performed the duty imposed upon him by his peers, he carried considerable money to Boston donated by:

Israel Fearing	10 shilling
Mr. John Elas (Ellis)	05 shilling
Mr. Joshua Gibbs	05 shilling
Mr. Samuel Burg	05 shilling
Mr. Thomas Bates	05 shilling

Mr. Ebenezer Berse (Besse)	04 shilling
Mr. Ebenezer Swift (the promise of)	05 shilling
Mr. Uriah Savery	05 shilling
Mr. Jirah Swift	05 shilling
Mr. Micah Gibbs	05 shilling

As Daisy Lovell, author of *Glimpses of Early Wareham* wrote — "it is well for us of this opulent and inflated age to pause and think how valuable and important money was in 1739." Daisy tells us in her research that "a shilling in relation to the spanish silver dollar was 1/6 of the Spanish dollar valuation or 16 2/3¢." This standard stood for more than a hundred years in New England.

Only a few men were involved in town government and it is apparent the offices were rotated from year to year since the names were always the same: Fearings, Bumpuses, Saverys, Swifts, Chubbuck, and Burges, as well as names of the original proprietors of the Agawam Purchase and other prosperous and influential people.

The Fearing family continued to live in their family house for several generations, making few alterations. The Fearing families eventually failed to keep the building in repair. A barn and shed used to shelter and feed the stage coach horses were allowed to deteriorate and finally collapse, leaving only a connected shed to the main house, both having been in poor condition prior to their end.

Mary Fearing, the last descendant of the Fearing family, who lived in the house, married William Warr, who mortgaged the property to construct a garage and storage building. During a difficult time the bank foreclosed the mortgage. Clyde Aiken purchased the property, installed his family in the Fearing home and bought a diner which he located on the north side of the property, on the south side of Route 28 passing the property on its way to Buzzards Bay and the Cape.

Mr. Aiken operated the diner until his death. The property was purchased by Mr. Ernest Blanchard, who then sold the Fearing homestead once known as the Fearing Tavern to the Wareham Historical Society for "one dollar and other considerations", in essence a gift to the Society through the efforts of Raymond A. Rider, the Wareham Historical Society president at that time.

Through the continuing efforts of the Society, spurred on by gifts from Dr. and Mrs. Engel, Mr. and Mrs. T. Rowe Price, some other Fearing heirs and some local philanthropists, some $50,000 had been raised and spent.

The Fearing Families in Wareham begin with: Israel Fearing, wife Martha Gibbs, and children:

Israel 2	Born May 1, 1723
John	Born Jan. 11, 1724
Benjamin	Born Sep. 21, 1726
Ann	Born Mar. 16, 1728
Noah	Born Feb. 10, 1731
David	Born Oct. 21, 1733
Elizabeth	Born Dec. 1, 1736
Martha	Born Mar. 29, 1740

The following children are attributed to Israel 1 but the last birth would make Israel 109 years old and is not consistent with the known date of Israel's 1 death in 1754 at the age of 72 years old; therefore it is thought these must be the children of Israel 2 above, although this Israel 2 would be sixty-eight years old, and would need a second wife, his first being Lucy whom he married (?) about 1767. None of this works out very well. We must conclude that a wrong entry was made, but we can assume very strongly these children did belong to a Fearing as they are entered under the name of Fearing and the given names are definitely of Fearing origin.

Further evidence is noted in Israel's 2 will when he writes, "As to my beloved wife Martha I leave her to her lawful Dower on thirds." She died in the same year Israel 1 died, 1754. He died in June, she in September of that year.

More speculation is perhaps fruitless, but it is more than probable that Noah Fearing, born in 1731, was the town clerk who recorded the family below, his own. He probably thought that everyone knew he was the father and saw no reason to enter his name.

The children of Noah Fearing (?) and wife Lucy:

Noah	Born Nov. 28, 1769
William	Born June 6, 1771
Thomas	Born Oct. 2, 1773
Hannah	Born June 23, 1775
Thomas 2	Born March 1, 1777
Mary	Born May 22, 1781
Charles	Born Nov. 29, 1784
Israel	Born Aug. 6, 1786
Lucy	Born June 31, 1789
Andrew M.	Born May 15, 1791

These children came over a span of twenty two years, so that if his wife Lucy was 20 years old when married to him, she would be forty-two when the last child was born. This makes for good sense and probability.

The children of Silas Fearing and wife, Hannah:

Patty Cleasamond .... Born Feb. 12, 1785
Peyton Randolph .... Born Dec. 19, 1786
Marietta .... Born Oct. 17, 1788
Rosella .... Born Mar. 29, 1791
Franklin .... Born May 3, 1793

David Fearing and wife Hepzebah Tobey:

Hepzebah and Nabby .... Born Sept. 13, 1795
Dorcas .... Born Sept. 13, 1797
Gesviah .... Born Nov. 2, 1799
Hepzebah 2 .... Born Jan. 20, 1802
Masella .... Born Mar. 11, 1804
Benjamin Tobey .... Born May 7, 1806
David .... Born Nov. 19, 1809
Caroline .... Born Aug. 5, 1812

Moses Fearing and wife Heziah Briggs:

Heziah .... Born Oct. 22, 1788
Hope .... Born Apr. 5, 1799

David Fearing, son of Israel Fearing the Squire, and wife Heildah Cushman:

Israel (General Fearing) .... Born Jan. 9, 1759
Abigail .... Born Apr. 1, 1760

Noah Fearing, wife Mary:

Lucy .... Born Jan. 10, 1772

Noah Fearing and wife, Nancy Nye, of Sandwich:

Martha .... Born Nov. 26, 1757
Silas .... Born Nov. 11, 1759
Paul .... Born Feb. 28, 1762

Noah Fearing and wife, Nancy Nye, of Sandwich (continued):

Noah 1 .... Born July 25, 1764
Mary .... Born Apr. 6, 1766
Mary and Noah 2 .... Born May 13, 1767

Benjamin Haskell Fearing 2, wife Salome Pope, married 1783/4

Elisha Pope .... Born Oct. 11, 1785
Benjamin A. Jr. .... Born Sep. 13, 1789
Salome .... Born Jan. 9, 1789
Louisa .... Born Feb. 22, 1791
Polly .... Born Apr. 18, 1793

Benjamin 2 .... Born Sep. 21, 1795
Joanna .... Born Feb. 27, 1798
Joseph Warren .... Born Sep. 6, 1800
William Sturges .... Born Dec. 26, 1804

Squire John Fearing and wife, Dorcas Swift:

Elizabeth .... Born Jan. 2, 1759
David .... Born Jan. 10, 1761
John .... Born Sep. 26, 1763
Moses Swift .... Born Oct. 2, 1765

Israel Fearing Jr. and wife Elizabeth Thacher:

David .... Born Aug. 24, 1780
Genas .... Born Dec. 29, 1781
Lucinda .... Born Sep. 23, 1784
Martin .... Born Jan. 3, 1787
Isiah .... Born Sep. 15, 1789
Thacher .... Born May 26, 1792
Fear Crocker .... Born Oct. 12, 1794
Oliver .... Born Jan. 27, 1801

Israel Fearing and wife Hannah Swift:

Israel .... Born Aug. 31, 1749
Sarah .... Born Mar. 2, 1750

Benjamin Fearing, Tavern Keeper, and wife Mercy (Mary?) Haskell:

Charity .... Born Apr. 19, 1752
Benjamin Haskell .... Born Oct. 31, 1755

Squire John Fearing (son of Squire Israel Fearing) and wife Nabby Gibbs. (The title now just honorary.)

Hannah Gibbs .... Born Nov. 2, 1792
Elizabeth .... Born Aug. 18, 1795
John .... Born Apr. 3, 1798
x Bethiah Gibbs .... Born Aug. 3, 1801
(married Abel D. Breed)
George Gibbs .... Born Oct. 3, 1804
x Andrew Coatsworth .... Born Oct. 3, 1804
(twins)
x Elbridge Gerry .... Born July 12, 1808
Albert Smith .... Born July 12, 1808
(twins)

Fearing Families

Elbridge Gerry .... wife — Ann Sophie Bolles
children

x separate stories on each individual

Andrew Gerry .... Born Jan. 5, 1833
John Fearing .... Born Oct. 20, 1837

children of

William Fearing and wife Elizabeth Nye

Henry Seymore .... Born Sep. 11, 1800
William Sidney .... Born Oct. 12, 1802
Daniel Butler .... Born Sep. 9, 1804
Lucy Maria .... Born Aug. 29, 1807
Hannah Nye .... Born Feb. 9, 1810
Charles Nye .... Born Mar. 10, 1812
Elizabeth .... Born Oct. 20, 1814

William H. and Delia Wing Fearing:

children

Charles Warren .... Born Apr. 22, 1861
Angenette Doggett .... Born Sep. 1, 1859
Flora Mary .... Born June 13, 1867
George Gibbs .... Born Oct. 6, 1869
William H. Jr. .... Born — no date given — 1871?

Birth of children of Alien Makepeace and W. Drisella Fearing:

John Fearing .... Born August 18, 1817
Died August 13, 1828
George .... Born Sept. 20, 1828
Died Sept. 26, 1829

John Fearing, Jr. and
1 child of two survived:

Noel Fearing .... Born November 23, 1821
Sarah Elizabeth .... Born April
Died April 20, 1820

Israel Fearing Jr. and wife Betsey:

children

Alfred A. .... Born October 7, 1807
Clarissa Ann .... Born March 4, 1809
Noah .... Born May 15, 1811
Israel .... Born Dec. 3, 1817
Mary Elizabeth .... Born August 8, 1822

Benjamin A. Fearing 3rd, Jr. married age 34?, wife Joanna Gibbs:

children

Benjamin .... Born November 5, 1824
Elisha Pope .... Born February 17, 1827
Mary Ann .... Born February 2, 1829
Francis Henry .... Born December 19, 1830

Frederick F. .... Born January 14, 1833
Elisha Pope .... Born November 30, 1834
Hannah F. .... Born October 5, 1836
Daniel W. .... Born December 16, 1837
Joseph E. T. .... Born July 22, 1840

Fearing Families in Wareham —

William S. and wife Hannah C. Barrows:

children

Elisha P. .... Born November 25, 1831
William H. .... Born December 22, 1835
Charles W. .... Born May 26, 1839

From Records of A. C. Fearing
Doan Street, Boston, Mass. 1876

"Sir" John Fearing came to America in the ship *Diligent* in 1638. He was a bachelor and settled in Hingham, Mass. He later married Margaret Hawkes. He is buried in Hingham.

Brig. Gen'l Israel Fearing of Wareham married in Sandwich to Lucy Bourne (daughter of Thomas and Lucy Bourne fifth generation from Richard Bourne) on March 2, 1769. Children were:

Noah .... Born November 28, 1769
William Sedley .... Born June 6, 1771
Thomas I .... Born October 2, 1773
Hannah .... Born June 23, 1775
Thomas II .... Born March 1, 1777
Mary .... Born May 22, 1781
Charles .... Born November 29, 1784
Israel .... Born August 6, 1786
Lucy .... Born June 30, 1789
Andrew M. .... Born May 15, 1791

Capt. William Sedley Fearing of Wareham, married in Sandwich Elizabeth Nye of Sandwich, May 25, 1799. Children:

Henry Seymore .... Born September 11, 1800
William Sedley .... Born October 12, 1802
Daniel Butler .... Born September 9, 1804
Lucy Maria .... Born August 29, 1807
Hannah Nye .... Born February 9, 1810
Charles Nye .... Born March 10, 1812
Elizabeth .... Born October 12, 1814

Lucy Maria Fearing married Perez Fobes Doggett, M.D. on November 26, 1832. Children:

William Sedley

Charles Seymore
Anna

William Sedley Doggett married June 21, 1866, Frances P. Willson. Children:

Lucy Fearing
Anna Frances
Amy Willson

Lucy Fearing Doggett married October 15, 1895 Ernest Joseph Silberberg. Children:

Gwendoln Frances
Alfred Fearing
William Sedley
Ann Violet

Edward Leopold (changed last name to Fearing 8-22-29)

Edward Leopold Fearing married in Winthrop, Mass., Edith Goodall Huby, September 5, 1933. Children:

Charles Edward .... Born May 6, 1934
Robert Ernest .... Born August 23, 1945

Charles Edward Fearing married Verna Valenta. Children:

William Joseph
Dana Charles .... Born July 2, 1957

Dr. Perez F. Doggett was a physician in Wareham and lived at the corner of Sawyer and Main Streets. The original home was later incorporated in the Eldredge Apartment building.

FEARING MARRIAGES AND OTHERS

Rev. Rowland Thacher — Abigail Crocker .... May 31, 1740
Israel Fearing — Hannah Swift .... April 28, 1746
Benjamin Fearing — Mercy Haskell .... Nov. 25, 1749
Rev. Rowland Thacher — Hannah Fearing .... July 20, 1754
2nd marriage
Noah Fearing — Mercy Nye .... Oct. 25, 1755
David Lincoln — Elizabeth Fearing .... June 2, 1760
Israel Fearing — Lucy Bourne .... Oct. 15, 1768
Andrew Mackie — Charity Fearing .... Oct. 26, 1770
daughter of Inn Keeper Benjamin Fearing
Rev. Josiah Cotton — Lydia Parker .... May 16, 1776
Israel Fearing — Elizabeth Thacher .... Oct. 9, 1799
Silas Fearing — Hannah Pope .... March 1783
Benjamin Fearing — Salome Pope .... Sept. 25, 1784
Moses Swift Fearing — Heziah Briggs .... Feb. 12, 1785
John Fearing — Abigail Gibbs .... Apr. 23, 1791
David Fearing — Hepzibah Tobey .... Feb. 20, 1793

Wm. Fearing — Elizabeth Nye .... Apr. 20, 1799
Jonathan Gilmore A.B. — Huldah Kendrick .... June 26, 1802
daughter of Capt. John Kendrick
Israel Fearing, Jr. — Elizabeth Gibbs .... Oct. 12, 1806
David Nye — Lucy Fearing .... May 27, 1810
Nathaniel Shiverick — Salome Fearing .... July 7, 1811
Peter Thacher — Elizabeth Fearing .... Nov. 17, 1815
Wm. Savery — Abigail Fearing .... May 28, 1817
John Fearing, Jr. — Drisella Swift .... Mar. 7, 1819
(8 years of marriages not in town records)
George Christie — Zeraiah Fearing .... Aug. 27, 1820
Dr. Eliphalet W. Hervey — Dorcas Fearing .... Aug. 14, 1825
Alvin Makepeace — Drusilla Fearing .... July 2, 1826
James Hervey, Jr. — Morilla Fearing .... Jan. 22, 1832
Abel D. Breed — Bethiah G. Fearing .... Aug. 12, 1832
Urill A. Murdock — Hannah N. Fearing .... Nov. 25, 1832
Capt. James C. Luce — Elizabeth Fearing .... Oct. 21, 1838
Elbridge G. Fearing — Abigail Bolles .... July 25, 1840
Nathan Lincoln — Martha Fearing .... Oct. 13, 1763
Barezillia Swift — Sarah Fearing .... Dec. 21, 1769
David Lincoln, Jr. — Elizabeth Fearing .... Sep. 16, 1760
John Fearing — Abigail Gibbs .... Dec. 22, 1791
Joshua Tobey — Hannah Fearing .... Mar. 19, 1796
Melateab Bourne — Mary B. Fearing .... Jan. 25, 1802
Curtis Tobey — Tirzah Fearing .... Dec. 27, 1804
David Nye, Jr. — Lucy Fearing .... June 14, 1810
Peter Thacher — Elizabeth Fearing .... Dec. 19, 1815
Wm. Savery — Abigail T. Fearing .... Oct. 5, 1817
E. Nye Chaddock — Hannah G. Fearing .... Oct. 3, 1817
(Chubbuck)
William S. Fearing — Hannah C. Barrows .... Nov. 11, 1829
Dr. Perez F. Doggett — Lucy M. Fearing .... Dec. 13, 1832
Rodolph W. Crocker — Sarah E. Fearing .... Oct. 14, 1839
Noah Fearing — Mary Nye .... Jan. 22, 1756
John Fearing Esq. — Dorcas Swift .... Mar. 23, 1758
David Fearing — Huldah Cushman .... Apr. 17, 1756
Israil Fearing — Hannah Swift .... Apr. 28, 1746
Israil Fearing, Jr. — Lucy Bourne .... Oct. 15, 1768
Miss Lucy Maria Fearing — Willis B. Chubbuck .... Dec. 20, 1832
Hannah N. Fearing — Urial A. Murdock .... Dec. 13, 1832
Elizabeth Fearing — Capt. James C. Luce .... Nov. 6, 1838
Sarah Elizabeth Fearing — Rudolphus W. Crocker

INTENSION OF MARRIAGE

Capt. Alden Besse — Mrs. Betsy S. Besse .... Dec. 25, 1852

ADDITIONAL MARRIAGES

Noah Fearing — Mary Nye of Sandwich .... Jan. 22, 1756
Benjamin Fearing — Salome Pope .... Nov. 25, 1784
David Fearing — Hepsibah Tobey Fearing .... Jan. 16, 1794
Noah Fearing — Lois Pope .... May 24, 1795
Israel Fearing, Jr. — Hannah Swift of Sandwich .... Nov. 13, 1746
Israel Fearing — Lucy Bourne of Sandwich .... Mar. 2, 1769
John Fearing, Esq. — Dorcas Swift .... Mar. 23, 1758
Capt. William Fearing — .... May 25, 1799
Benjamin Fearing — Mary Haskell .... Jan. 11, 1749
Israil Fearing — Elizabeth Thacher 1799
Moses Fearing — Thesiah Briggs Fearing 1786
Silas Fearing — Hannah Pope .... Mar. 1783
John Fearing, Jr. — Drusella Swift .... Mar. 7, 1819
William H. Fearing — Deliah Wing 1857
John Fearing — Nabby Gibbs 1790
Elizabeth Fearing — Ann Sophie Bolles 1831
Andrew Coatsworth — Alderveronicafloscoraphonia
their daughter Elizabeth, M. William Root Bliss
Elisha Rider — Rebecca Hall, both of Middleboro 1807
Benjamin Tobey of Boston — Sally Swift of Wareham 1809
Peter Mackie (Dr.) — Fanny Nye of Sandwich .... Feb. 18, 1813
Capt. Stevens Gibbs — Miss Louisa Fearing .... Nov. 5, 1817
Rufus Lincoln — Mercy Stevens of Fairhaven .... Nov. 3, 1817
Bradford Lincoln — Mercy Gibbs .... Feb. 29, 1816
John Fearing, Jr. — Druscilla Swift 1819
Zerviah Fearing — George Christi .... Sep. 20, 1820
Sylvanus Bourne — Hannah Smith .... Aug. 23, 1823
Doctor Eliphalet Hering — Dorcas Fearing .... Aug. 25, 1825

DEATHS BOOK A.

Israel Fearing Jr. .... died September 2, 1753
Israel Fearing, Esq. .... died June 9, 1754
Martha Fearing(wife of Israel, Esq.) .... died September 9, 1754
David Fearing .... died April 27, 1760
Bethia Fearing .... died April 5, 1761
Noah Fearing .... died October 8, 1764
Mary Fearing .... died 1767
Noah Fearing .... died August 28, 1767
Thomas Fearing .... died December 29, 1774

Thomas Fearing 2nd .... died October 3, 1778
David Fearing .... died September 25, 1780
Charles Fearing .... died July 16, 1786
Hope Fearing .... died Novmeber 6, 1791
Thatcher Fearing .... died July 4, 1794
Hepzibah Fearing .... died April 20, 1796
Hepzibah Fearing 2 .... died November 22, 1804
Alfred A. Fearing .... died January 26, 1809
Dorcas Fearing .... died April 18, 1820
Noel Fearing .... died April 20, 1820
Elisha Fearing .... died March 20, 1822
Hannah F. Fearing .... died February 14, 1837
Migail Fearing
died at the age of 73
Betsey Fearing .... died at the age of 90 — November 5, 1848
George G. Fearing .... died at the age of 5 yrs, 5 mo. Dec. 17,

FEARINGS BURIED IN CENTER CEMETERY

Eliasha R., M.D. .... died 1876
Mary Ann .... died 1852
Benjamin Fearing .... died January 5, 1866
Joanne Gibbs Fearing .... died July 26, 1878
Mary Ann Coggeshall .... died December 5, 1921
Francis H. Fearing .... died October 15, 1907
George Fearing .... died August 22, 1879
Flora M. Fearing .... died July 30, 1869
Charles W. Fearing .... died April 9, 1863
Mrs. William Fearing .... died March 27, 1916
William Fearing (date in record book missing)
Charles W. Fearing .... died November 16, 1845
Elisha P. Fearing .... died 1863
William S. Fearing .... died February 25, 1862
Elbridge G. Fearing .... died July 16, 1894
Ann S. Fearing .... died December 29, 1839
Abigail B. Fearing .... died October 29, 1894
George G. Fearing .... died December 17, 1830
Ellie G. Fearing .... died August 15, 1860
Ella Fearing .... died January 13, 1869
Albert S. Fearing .... died February 18, 1836
Lena S. Fearing .... died April 9, 1930
William Sedley Fearing (no date)

CHAPTER XI

From Isaac Bumpus to Israel Fearing

Rochester Town, earlier known as "Sippican Proprietary", was owned by nearly thirty individual proprietors. Israel Fearing was not one of these early proprieters, but by grant obtained large acreage in the area.

The names of the Rochester Proprietors are:

Joseph Lothrop	George Morton	John Perry
Barnabas Lothrop	William Dexter	Samuel Hammond
Candum Winslow	Samuel Briggs	Benjamin Hosten
William Clarke	Seth Pope	Benjamin Bartlett
Ralph Bradford	Samuel White	Elizabeth Ellis
Ralso Powell	Joseph Doty	Joseph Dunham
Samuel Arnold	Will Connit	Thomas Hinkley
John Burge	Francis Clarke	Thomas Clarke
Joseph Burge	Aaron Barlow	William Paybody
John Cotton	John Bradsard	

Just south of the Tihonet lands on the east end bordering the Wankinco River lies the land of Isaac Bump, together with that of John, Edward and Jeremiah Bump comprising the four sea lots then in Rochester Proprietary. This land was laid out in 1685. More of this area was laid out on April 17, 1740 and was acquired by Edward Bump, being the large lot and small lot on which his house still stands (now divided into four apartments on a hill overlooking a shopping center). The Bumpus land contained an area extending from the high dam area down the Wankinco River's west side through Pinehurst and bounded by Broad Marsh River extending to and including all the meadows along Broad Marsh Creek to what was called "the head line" of the neck or about 80 rods above the dam area. This whole area was called "Wonkinco Neck."

On 7 April 1747, Isaac Bumpus sold to Israel Fearing all of his lands, buildings and water rights as recorded in a deed of that date. Israel Fearing was then sixty-five years old. He died in 1754 leaving this whole property complex to his son Benjamin, stating Benjamin was to provide rertain areas of the dwelling house and barn for use of Hannah Fearing, widow of Israel's son, Israel, so long as she remains his widow "and no longer." Younger Israel died in 1753.

The Deed to Israel Fearing

To all people to whom these presents shall come, Greetings. Know ye that I, Isaac Bump of Wareham in the County of Plymouth, in the province of the Massachusetts Bay in New England, Yoeman, and in consideration of the sum of two thousand pounds in bills of the old tennor to me in hand before the ensealing hereof well and truly paid by Israel Fearing of the same town, county and province aforesaid. Gentlemen the receipt whereof I do hereby acknowledge and myself therewith fully satisfied and considered and thereof and of every part and parcel thereof do exencrate, acquit and discharge him the said Israel Fearing and to his heirs, Executors and Administrators forever by the presents. Have given, granted, Bargained, sold, aliened, conveyed and confirmed, and by these presents do freely, fully and absolutely give, grant, Bargain sell, aline, convey and confirm unto him the said Israel Fearing and to his heirs and assigns forever, all that my several pieces or parcels of upland and meadows, housing and lying in Wareham aforesaid, hereafter mentioned and my homestead in which my dwelling have now stand. It being the fourth sea Lot and also the third sea lot on which my barn stands, Also twenty acres of additional land lying up the Brook taken up by grant of the proprietors. Also a parcel of land lying in Wonkinco Neck, which fell to me in the division made with my brother as may be seen by the division deed. Also two pieces of additional land lying on the northerly side of the fourth sea lot, the one ten acres, the other thirteen acres as appears on deed. Also two acres lying on the west end of the third and fourth sea lots lately laid out. Also a parcel of land and salt meadow lying in Wankinco Neck, which I hold by deed under the hand and seal of my brother John Bump as may be seen by the deed. Also a parcel of land lying in the bounds of Plymouth which I hold by one deed under the hand and seal of Joseph King in which there is a cedar swamp. Also all my meadow lying on the west side of Wankinco River from the mill down to the lower end of the land I had of my brother John Bump.

Also a piece of salt meadow lying at the Broad Marsh which I had of Aaron Barlow all except what of it I have sold to my brother John Bump and also one last part of a share in the undivided in the Proprietary of Rochester together with my dwelling house, barn and other buildings thereon standing on land Fancy, and to have and to hold the said granted and Bargained herein for, with all the appurtenances to him the said Israel Fearing and his heirs and assigns forever to his and theirs only proper use Benepts and Behoof forever. And I the said Isaac Bump for me my heirs, Executors and Administrators do covenant, promise and grant to and with the said Israel Fearing his heirs and assigns that before the ensealing hereof I am the true, sole and lawful owner of the above bargained premises and lawfully seized and possessed of the same in my own right as a good perfect and absolute Estate of inheritance in Fee simple. And have in myself good right, full power and lawful authority to grant, bargain, sell, convey, consign said Bargained Premises in manner as aforesaid and that the said Israel Fearing, his heirs and assigns had and may have from time to time and at all times and forever hereafter by name and virtue of these presents, lawfully, peacefully and quietly have his use, occupy, possess and enjoy the said promised and Bargained premises with the appurtances free clear and freely and clearly acquited, Exonerated and discharged of from all and all manner of forms of other gifts grants bargained labor, leases, mortgage, wills, Intails, Jointures, Dowering, Judgements, executions, or Incumbrances of what nature or nature ever that might in any major degree, obstruct make valid this present deed. Furthermore I the said Isaac Bump for myself, my heirs executors and administrators do covenant and engage the above identified premises to him the said Israel Fearing his heirs and assigns against the lawful claims or demands of any person or persons whatever, forever, hereafter to warrant secure, defend by these presents. In witness whereof I the said Isaac Bump have hereunto set my hand and seal this 7th day of February anno Domini 1749th

Isaac Bump (seal)

and I Elizabeth Bump the wife of the above mentioned Isaac Bump do hereby give up my right of dowery and power of thirds of in and unto the said Israel Fearing and his heirs and assigns as witness my hand and seal, ye day of above Her Elizabeth Bump X (her mark and seal)

Isaac Bump was then 68 years old, son of John and Sarah (Hunter) Bump. He was one of twins with Zoar, born 27 March 1681.

Five years later Israel Fearing, Esquire died leaving a will bequeathing this entire estate to Benjamin his son, born 21 September 1726.

Squire Fearing's son Israel preceded him in his death by one year. To his widow Hannah he left the privilege of occupancy in Benjamin's dwelling plus other living rights so long as she remain his son Israel's widow "and no longer".

The Bump-Fearing deed successfully transferred all of Isaac Bump's Property in the four sea lots and elsewhere as listed in the deed, to Israel Fearing. Israel, as noted, soon bequeathed this property to Benjamin Fearing in 1754, who very soon thereafter opened Isaac's dwelling house as a house of entertainment or "Tavern". It became a central place for meetings of the selectmen, proprietors and a place for Benjamin's brother, Squire John Fearing, His Majesty's Justice of the Peace, to hold court for the transgressors of His Majesty's laws and local regulations, brought in by the Constable.

Eleven years from the date Benjamin inherited the property, he expanded the building by adding four rooms on the South side and turning the "leanter" into the tap room. Any obligation he had to his dead brother's wife had disappeared, as Hannah was soon married to the Rev. Rowland Thacher.

From a small dwelling house of four rooms, two downstairs and two on the second floor, the reconstruction made an imposing Inn or ordinary. The second floor bedrooms were elegant with their paneled walls, ornate closet-cabinets and fluted columns on each side of the small fireplace, wide floor boards and moulded window seats. Interior solid blinds kept out the sun's rays in summer and the cold in winter from the large windows.

The front was clapboarded, the sides and roof were shingled with "shakes", a hand split shingle cut carefully from cedar logs sawed at two foot lengths, then precisely cut lengthwise from selected straight-grained material.

The date 1765 was finger-drawn in the wet plaster, being a means of identifying the date and sometimes the initials of the workman while building the structure — the custom persists even today. This date can be clearly seen as you enter the south side main entrance looking to the east or right wall.

Two huge chimneys pierce the roof to provide draft for three fireplaces in one, and four fireplaces in the other. A Dutch cooking oven is in the Publik room and also in the tap room area.

About 1830 a wing was built on the north side with a door

connecting it to the Publik room. Three rooms on the second floor were probably reserved for the immediate family.

Israel Fearing's will (he was born August 29, 1682, and died July 16, 1754) gives Benjamin Fearing all of the property he purchased from Isaac Bumpus, all the buildings including the grist mill he bought of Peter Oliver, the house and lands he bought of John Norris, the lands and swamps he bought of Samuel Savery, land and meadows at Little Neck pine neck, land he bought of Thomas Tupper, land bought of Zaccheus Bumpus, and 1/5 part of the Cedar Swamp on the west side of Wankinco River.

He provides for his son's widow certain land and privileges in the home of Benjamin "so long as she shall remain a widow."

John was Israel's third son, but he must have been his favorite because he was the first son to be named in his will after Israel's wife Martha. She was to have her dower of one third of the estate as the will stipulates: "I leave her to her lawful dower or third."

The real estate was worth 4000 pounds, the balance of the estate as listed was valued at 510 pounds, making a total of 4510 pounds, an enormous sum in those early days.

LAST WILL OF ISRAEL FEARING 1754

July 2. Day 1754
An Inventory of the Estate of Israel Fearing Esqr. late of Wareham Deceased . . .

To Armes & Ammunition 4. O. 1 To Wareing apparel		15.18.8
To Books (lbs) 4.5.0 To a hone & House lock & Sundry Other things	15.10.0	17.18.8
To case bottles 13/4 to Beding & Furniture (lbs) 57.12-0		58.6.0
To Chist Draws & Chists & Sundry Other things		55.0.2
To Hogsheads & Sundry Other things		0.5.8
To Hollow Ware, Old Iron & Sundry Other things		6.14.0
To Loombs		2.13
To puter & knives & Forks & Sundry Other things		15.8
To Money & Notes		225.9.0
To pockett Compas & Sundry Other things		(illegible)
To Chains Axes and Sundry Other Small things		3.9.0
To Corn & Rye, and Chirn & Sundry Other things		2.12.3
To 3 Spaids, & Sundry Other Small things		15.10
To 2 Mairs & Coalt & 4 Cows & 2 Calves		33.3.4
To Hay & Solks(?)		1.17.0
To Shingles & Boards & Sundry Other things		4.6.4

To Corn & Rye & Wheat and flax seed	(lb)	0.4.0
To Meat and fat and Cheese		3.8.0
To and Sundry Other Small things		.15.
To 2 Carts & Tackling for Said Carts		6.3.1
To 2 Saddles and Sundry Other small things		1.18.0
To apair of Oxen		0.6.0
To the Raell Estate	(lb.)	4040.4
		4510..10..3

Plimoth July 6th, 1754
Noah Fearing (illegible)
(illegible) Execr. to the said
Ebenr. Bourge
Thos. Bates
David Besse

Decd. Will, made Oath, that this Inventory contains all the Testators Estates so far as they know of, and if more hereas appears theyl. give it in the said apprizers being also under Oath at the Same Time . . . .

Before John Cushing Judge of Probate

In the name of God, Amen, this 23d. day of Janry., 1754.

I, Israel Fering of Wareham in the of Plimouth and province of the Massachusets Bay in New England Gent. being aged and Infirm, yet of perfict mind and memory thanks to God, Do make and Ordain this my last Will & Testament; Respecting such Worldly Estate wherwith God hath Blessed me in this life, I Do Give, Demise and Dispose of the same as follows. Vivr.

Item My Will is that all my Just Debts and Funeral Expenses be first paid by my Executor hereafter nam'd . . . .

Item As to my Beloved Wife Martha I Leave her to her lawful Dower or thirds . . . . .

Item I Give and Bequeath to my son John Fering to him, his heirs & assigns forever, the following Real Estate. Vizt. apiece of land & meadow at Bass Creek, so called, Extending Easterly to the Land I bought of Barnabas Bates, and to the meadow I Bought of Thos. Tupper, & Westerly to the lands of Joshua Gibbs; also Another peice of Land & Meadow called Crooked River, Bounded as follows. Westerly by the lands of Ebenezer Bourn, Southerly by the Roade Northerly by Warren's lands; also apiece lying between the two Roads which lead to the necks Called Indian Neck & Great Neck bounded Easterly by Bourge's land, Southerly by Bourn's lands; Also Fifty Acres Lying by the Flaxpond above the Roade which Leads to Sandwich; also apiece of Meadow Called Miery Meadow lying at the River Bounded beginning at the water fence by the River and Runing Down the River till it comes to a small white Oak tree near Goose Island, and from (that) tree to the River including Goose Island, with apeice of land lying by said Meadow, Extending the length of said Meadow about four Acres to be Lett of for Conveniancy of Fencing the Meadow; also apeice of Land in the pine Neck being my Right in the fourth lott. Excepting what I bought of Joseph Gifford also One Fifth of my upper Cedere Swamp lying On the West Side of Wonkimco River. Also One fifth part of my Grist Mill Stream and of my Low lands at and Naer Rowland Swifts also a fourth part of my Low lands & Swamps lying above the Uper Roade, which lands from the House of David Bessey to Sandwich, also my Dwelling House at Red Brook; In Case he move of the same within Two years after my Decease, also One half of the Frute Groing On my Orchard at the little Barn During Ten years, it being the Southerly half . . .

Item I Give and Bequeath to my son Benjamin Fering, to him and his heirs assigns forever, the Real Estate hereafter named, Vixt., all the lands Meadows and Buildings which I Bought of Isaac Bumpas (Excepting as is hereafter Excerptd.) Also my Lands and Grist mill which I bought of Peter Oliver Esqr. Also the House and Lands I bought of John Norris, also the lands & swamps I bought of Samuel Savery, also a lott I bought of Ebenezer Perry, also a peice of Land and Meadow Called Litlle Neck, Bounded Northerly by Bourges's land and from Bourgas's Corners through a small pond into head of Crab Coce, also an Island Called Tim's Island, also apeice of Meadow in pine Neck, which I bought of Thos. Tupper Adjoyning to the second and third Lotts of Land, also the land I Bought of Zacheus Bumpas, also One fifth part of the upper Cedarswam Swamp lying On the West Side of Wonkimco River, I also Give and alow Said Benjamin his heirs & Assigns the following Improvements & priviledges which he & they are to Enjoy so long as Hannah Fering Widow of My son Israel shall Continue his Widdow and not afterward, Vizt. he and they shall have Ten Coard of Oak Wood yearly off of the upper End of the land called

Warrens Point, also the priviledge of pastureing two Cows Dureing the Season of the pastureing yearly; partly in Indian Neck and partly in Crab Cove, also One half of the frute which yearly shall Gro in the Orchard I bought of Jonathan Chubbock, also the yearly Use of a peice of Salt Meadow Joyning to the West End of the Meadow which was Thos. Tuppers, Bounded Runing up the Crick till it Comes Over Against a Marked Tree. Bounded Easterly by the meadow of Barnabas Bates & what I bought of John Bates, also One third of the Great Flat being the Easterly End, to Improve as aforesaid . . . .

Item I Give and Bequeath to the Children of my son Israel deced., vizt. Israel and Sarah, as follows to Israel Fering, Jr. his heirs and assigns forever the Following Reail Estate, Vizt. all my lands and Meadows in the Indian Neck which I have not herein before Given Away, also the Homestead which I bought of Jonathan Chubbock, Bounded On the Westerly End by Warren land, and by Roads Easterly till it comes to the Brook, and so to the head of the Brook to a marked white Oak tree thence Runing to apond near the path that Leads to the River and so through the Middle of the pond to a stump with stones, and so to the narrow place of the Sampear (spelling?) Meadow, and a straight Range to Benjamin Norrises pint & into the River, also All that I Own to the Westward of the lein last Nam'd, Excepting what I have before Given away, also apeice of Meadow On the lower side of the Brook Adjoyning to Bates & Warren, Including the Swamp, also Another part of my land in the long neck, which lyeth On the westward Side of the Ministerial lands, also apeice of Meadow in the pine Neck, Bounded by the Meadow, and head of John Bates and by the Meadow of Barnabas Bates Runing bounded to a mark'd pine tree to the nearest place of the Crick, also apeice of Meadow Called the Great Flat being One half at the Easterly End with apeice of Land two Rodds in Wedth Runing along by the Meadow to the Ceder Tree and Runing Round taking in all the Meadow and the same Wedth of Upland to a pine Mark'd with stones, near the Island and thence Through the Middl of the Great Island to a tree mark'd with stones and so Into the water Against the Meadow I had of Gifford Including apeice of Land Lying by the aforesd. Meadow of Barnabas Bates Including two Acres of Land by the Meadow of Conveniancy of Fencing, also a peice of land in pine Neck being the second Lott at the Westerly End, also Excepting the priviledges I have Givne to my son Benja and my daughter in law Hannah Fering . . .

Item I Give to my said Grandaughter Sarah a Legacy of one hundred and thirty three pounds six shillings & Eight pence lawfull money which my Grandson Israel shall pay sd. Sarah when she comes of age . . . .

Item I Give and allow to my Daughter in law Hannah Fering Dureing her Widdowhood the following priviledges, namely the use of the Westerly lower Room in the Dwelling House I have Given to my son Benjamin, with the Westerly Garrit One third of the seller, One third of the Leanter at the west End, with liberty to Bake & wash in said Leanter also pastureage for two Cows yearly in the season of pasturing to Go with Sd. Benjas. Cows with the priviledge of useing One third of the little Barn On said Farm also Liberty to Improve asmall field about half an Acre On the south side of the Roade near sd. House now fenced in, also liberty of Cutting Ten Coard of Wood yearly off of the Lott I bought of Ebenezer Perry, also One half of the Frute yearly which shall Gro in the Orchard by said Dwelling House, also Pasturage and hay for ten Sheep yearly On the Farm I have Given to her son Israel Only in the Spring Fall & Winter, also Fodder or Salt hay to winter Two Cows yearly off, of the place I have Given to her son Isreal . . . .

Item I Give to my son Noah Fering to him his heir sand assigns forever, the Real Estate hereafter named, Vizt. apeice of land On the Northerly Side of popes pond Bounded by Bourges's And the Road, also apeice of upland and Meadow lying On & Adjoyning to the East Side of Samuel Bourges's land & Meadow, Bounded begining at the Crick & Bound by sd. Bourge untill it Comes to the Roade which Goes to Sandwich, and so by said Road, to the Land of Micah Gibbs and Southerly by said Gibbs till it comes to the Muddoy Cove and by sand Cove and the Salt water Southerly & Westerly to the First Mentioned Bounds, Excepting Reserveing from said Noah, what I have or shall herein Give to my Other Children, also apeice of Salt meadow On the southward Side of the Crick in the Broad Mash, Being apeice I bought of Thomas Tupper, Bounded by Burges and by the Meadows which I bought of Joseph Gifford, also two peices of land I bought of Barnabas Bates, Bounded partly by Thos. Bailey and partly Burges Land Including, the peice of Meadow within the Land which I bought of Thomas Tupper, also apeice of upland Lying in long Neck On the Westward side of the ministree lands, being One half of my Right there being the south ward End also apeice of Meadows which I bought of Samuel Bates lying On the southward Side of Eggawam River Bounded partly by the Meadow of Thomas Bates, and Runing up the River till it Meats with the Meadow which I have Given to my son John Takeing in astrip of upland ten Rods in Wedth, and runing the Length of Said Meadow.

Item I Give and Bequeath unto my son David his heirs & assigns forever the Rael Estate now to be Named, Vizt., my now Dwelling House and Barn and Other Buildings Standing On the land now to be Given to him, and the House I Bought of Samuel Bates, also lands & Meadows, as follows apeice by my house Bounded Easterly By Bourge's land Runing North to the River, and Down said River till it Comes to the Meadow and Land which I have Given to my Grandson Israel, and Runing by him Southerly till it Comes to the Road & by the Road to the first mentioned Bounds, Excepting but what I have herein before Given to my son John & Noah, within these Bounds. Also the Westerly half of apeice of Meadow called the Great Flat, Including two Acres of land Lying by the side of said Meadow for Conveniancy of Fencing, also apeice of Meadow lying in Wilbors Neck sufficiant to Cut Two Load of Hay yearly to be sett off To the Northward of the Great Island, and Eastward of the small Island, also apeice of Meadow lying in the Broad Mash which I bought of Joseph Gifford, Also another peice bought of said Gifford last, being salt meadow lying in pine neck Joyning to the second Lott of upland in said Neck and Lying Bound asmall Island also anoth peice of Land and Meadow Called Onset, also Another peice of land lying in long Neck On the Eastward Side Of the Ministerall lands being all my Right there also another peice of land lying On the Westward side of said Ministerall land, being a Quarter part of my Right there, also a quarter of a lot of upland Lying in pine Neck, being what I bought of Joseph Gifford, also another peice of Meadow lying in the Broad Mash East Side of the Crick and the upland On Each side of the Meadow not being herein before Disposed off . . . .

Item I Give to my sons Noah & David and to my Grandson Israel and to Each of them their heirs and assigns forever the Following lands and Swamps, Vizt. to Each One afifth part of my upper Ceder Swamp lying On the West side of Wonkimco River also to Eich of them a Quarter part of the Gristmill by Roland Swifts with a Quarter of the stream & low land belonging to said Mill, also aquarter of my Out lands and Swamps which Lyes above the uper Roade which leads from the Dwelling House of David Besse to Sandwich. To them their heirs & assigns forever, Excepting what I have Given to John . . . .

Item I Grant and Allow to Each of my sons their heirs & assigns liberty of all necessary and conveniant ways for passing and Repassing Over Each Others lands where it Shall be most Conveniant and Least prejudiciall to Each of the parties . . . .

Item I Allow to my Daughter so long as they shall Live Unmaried, liberty of the One half of the Dwelling House which I have Given to my son David, and he shall find and provide them fire Wood so long as they shall Dwell in said House. & I also allow them the same liberty in the West Chamer of the Dwelling House which I have Given to my son Benjamin, and he shall find and provide them fire Wood while they shall Dwell in his House . . . .

Item I Give and Bequeath to my three Daughter Ann, Eliza & Martha to Each of them a legacy of One Hundred and thirty three pounds six shillings and Eight pence lawfull money to be paid within two years after my Decease and my sons John, Benjamin, Noah and David shall pay to my said Daughters to Each of them Eighty pounds of Such money & my Grandson Israel Or his Guardian shall pay Eighty pounds Such money to my said Daughters . . . .

Item I Give to each of my said Daughters a Good Bed of Furniture

Item I Give to Each of my four Sons and three Daughters, & my Grandson Israel to Each of them Equally all my Household Goods and Utensils

Item I Give to my said Four sons and To my Grandson Israel Equally all the Rest of my personal or Moveable Estate

Item I allow my son John priviledge of Dwelling in the House I have Given to David so long as said John lives unmarried, and it shall be in the Easterly Chamber

Item I Order my son David to provide yearly pasturage for a Cow for my three Daughters, and my son Noah shall find fodder for such Cow and Look after sd. Cow between them, this Benefit my Daughters have no longer than they live Unmarried . . .

Item I Appoint my son Benjamin Guardian to the Estate which I have Given to my Grandson Israel to Raise money to pay the Legacy which is to be paid to my Daughter Out of said Estate, and to pay it in Due Season, and in Other matters to act According to Law and to be accordingly Accountable

Lastly, I Do Constitute and appoint my son Noah aforesd. to be sole Executor to this my last Will and Testament and I Do hereby make null and Void all other, former Wills and Testaments Bequests and Execrs. before named Willed or Bequested Ratifying & Confirming this & no Other to be my last Will and Testament, In Witness whereof I have hereunto sett my hand and Seal the Day and year First above Written Israel Fering (seal)

Signed Seal'd, Deliver, published
Pronounced & Deilard (?) by said Israel
Fering Esqr. as his Last Will & Testament
On the presents of us Witnesses,

Ebenezer Burge
Barnabas Bates
Elisha Tupper

Plimo. St, July 16, 1754, This Will being Exhibited for proft. by the Execr. therein Named the said Ebenezer Burge & Barnabas Bates, made Ooath (that) they Saw the Testator Signs & Seal & heard him Declare this Instrumt, to be his Last Will & Testamt., and (that) they (together with said Elisha Tupper), at the same Time Subscribeds as Witnesses

Before Me John Cushing Judge of probt.

(SEAL)

Province of the Massachusetts Bay Plimouth Sc, By the Honble. John Cushing Esqr. Judge of the probates of Wills & for Granting Letters of Admn. on the Estates of person Decd. haveing Goods, Chattles, Rights and Credits in the County of Plimouth in the province aforesaid . . . .

To All unto whome these presents shall come Greetin Know ye that upon the Day of the Date hereof before me at Plimouth in the County aforesaid the Will of Israel Fering late of Wareham in the County of Plimouth Esqr. Deceas'd to these presents Annex'd was proved Approved & and Allow'd who haveing while he Lived and at the Time of his Death, Goods, Chattles, Rights and Credits in the County aforesaid, and the probate of the said Will & power of Commiting Admn. of all and singular the Goods, Chattles, Rights & Credits of the sd. Deceas'd by Virtue thereof appertaining unto, (illegible) the Admn. of all and Singular Goods, Chattles, Rights and Credits of the said Deceas'd & his Will in any Manner concerning is hereby Committed unto his son Noah Fering of said Wareham Executor in the Same Will named well & Faithfully to Execute the said Will, & to Admn. the Estate of the sd. Decd. According thereunto and to make a True and perfict Inventory of all and singular the Goods, Chattles, Rights & Credits of the said Deceas'd, and to Exhibite the same into the Regestry of the Court of probate for the County aforesd. at or before the 16th. Day of October next Ensuing, and also to Render aplain and True account of his sd. Adm. upon Oath within Twelve months next Coming, In Testimony Whereof I have hereunto Sett my hand and Seal the 16th day of July Anno, Domini, 1754 . . . John Cushing

General Israel Fearing Homestead

Fearing Tavern as a private home about 1850

CHAPTER XII

Benjamin Fearing Estate

The estate that Benjamin the first inherited from his father, Israel, was a huge and complicated property. In the dam area several industrial works had been operated; namely a grist mill, a fulling mill, a saw mill and a small iron works.

Reverend Rowland Thacher, who built his house on the south side of the old country road directly in the path of the Old Colony railroad when it came to Wareham in 1848, was on land called the third sea lot. This land was bought from Isaac Bumpus by the town for the ministry's use.

Rev. Thacher's salary was always in arrears, sometimes as much as a year or more. The minister who married Israel Fearing's widow Hannah as his second wife, was then related to Benjamin Fearing by way of this marriage. Benjamin had inherited a fulling mill on the Wankinco River as part of the legacy from his father, Squire Israel Fearing, who married Martha.

Shortly after Benjamin acquired this diverse estate parson Everett rented the fulling mill from Benjamin and no doubt acquired an income from this mill, whose operation could be plainly heard from the meeting house on week days in the summer months. It was a noisy operation.

Acres of salt marsh, cedar, swamps and uplands with woodlands enabled the parson to make a good living as long as his health was to last. He built fences, cut wood and raised his own food and flax for weaving. He was also expected to visit the sick, performs at funerals and christenings, and bellow forth his sermons for several hours on each Sunday. He was sick only a short time before his death in 1775, and nearly lost his parish, but he died before he could be replaced by the town fathers, who could not tolerate paying for the minister who failed in his duties, no matter what the reason.

Parson Thatcher's house was twice moved and now rests on a lot facing Cranberry Highway or Route 28 — a modern highway in the early 1930's, now superceded by a new and more modern four lane highway a half mile north of Route 28.

The Wareham Historical Society is slowly renovating and restoring the old parsonage as funds become available. The old central chimney was lost in the first moving, as were all of the old fireplaces that went with it.

We must leave the story of the old parsonage for now as we go on with our story of the Fearings and Fearing Tavern.

Benjamin showed himself to be a good business man, proceeding to open the former Isaac Bumpus house as an inn. As stated, he rented to Parson Thacher the fulling mill and the iron works to the Pratts and Wilbur. In the agreement with the Pratts and Wilbur he prevents them from operating a grist mill or a house of entertainment (this would have been in competition with Benjamin and he was not about to have any, at least no so close to his own enterprises).

Although there was a saw mill there at the time Benjamin Fearing inherited the property, there is no mention of its continued operation. One suspects that this operation ceased. It is probably true that the Pratts and Wilbur converted this operation into a nail factory, since the powerful water wheel and machinery were available for conversion. Several other buildings already existed, as indicated in the agreement. The saw mill and the dam were on the west side of the Weweantic River, as were the sheds and stave building.

There was also another grist mill on Bumpus Brook, now known as Rose Brook, under Benjamin Fearing's ownership.

Israel Fearing, Benjamin's father, was a shrewd man. He knew his son's capacities very well. To John, his third son, he left real estate and the title of Squire. John was worthy of this title as history records his faithful performance of this office, recorded in his "Book". John is the first mentioned in Israel's will. Benjamin was second in the will. Israel's son Israil's children received a house and land and the granddaughter money to be paid in her legal maturity. Thirdly to his son Noah, fourthly his son David who did not make his mark in history other than marrying Heildah Cushman and having two children, Israil and Abigail. This last Israil Fearing became famous as a general in the War of 1812.

Son John was unmarried at the time of his father's will and was permitted to live in the house that Israel had given to David as long as John remained unmarried. John married Dorcas Swift in 1758. He then moved to a house he had built on Rochester Road or County Road, now known as Fearing Hill Road, where he became the squire shortly after, and remained with this title until his death.

Elbridge Gerry Fearing inherited a large part of his father, John Fearing Esquire's, estate from which he was to pay all of the debts of the estate. Among the debtors was Abel D. Breed who had married Elbridge Gerry's sister Bethiah G. Fearing, August 12, 1832. Abel D. Breed held some notes amounting to almost three thousand dollars, according to an agreement October 31, 1846 copied May 15, 1877 which proposes to swap the notes for certain parcels of property inherited by Elbridge Gerry Fearing, including the Squire Fearing homestead and its contents. The will was executed November 9, 1845. One third of all the property inherited by Elbridge Gerry Fearing was deeded to Abel D. Breed by agreement of both parties. This deeded the Squire Fearing homestead to Abel D. Breed and Bethiah Fearing Breed, a sister of Elbridge Gerry Fearing.

Another Elbridge Gerry Fearing agreement is recorded in which it is stated:

> "To permit the widow Rebecca Barrows to use and enjoy one comfortable room and to have sufficient privilege of cooking in said house free of rent and to permit my brother Andrew C. Fearing and his wife to live and dwell on said premises free of rent for and during their joint lives and the life of said Andrew C. Fearing who can pass this property to his heirs.
>
> (signed) Elbridge G. Fearing
> (signed) Abigail Fearing (wife)
> (signed) (Abigail Bolles)

David inherited "my now dwelling house and barn and other buildings standing on the land now to be given to him, and the house I bought of Samuel Bates." From his father Israel Fearing, who died in 1754, David's son became General Israel Fearing, who occupied the dwelling house after his father's death.

This Israel Fearing's house was burned to the ground by vandals, but the Samuel Bates house still stands on a knoll overlooking the Great Marsh easterly and southerly of this house.

It was General Israil Fearing who rounded up the neighborhood men to oppose the British raid on Wareham from the bark *Nimrod,* a British gunship.

The British set fire to the cotton mill, burned on the ways several ships nearly ready to launch and took some hostages, preventing General Fearing and his men from taking any action. The hostages were dropped at Nobska Point as the British left to board the *Nimrod* and head for Newport, Rhode Island.

CHAPTER XIII

Some Bumpus Genealogy

Isaac Bump, who sold his entire property of mills, marshes, uplands, cedar swamps and water rights to Israel Fearing, was strictly a business man and never held a town office, although his brother Edward and Jeremiah were elected to several town positions. He was, however, the richest of all the Bumps. He took 'Old Tenner' in payment from his sale to Israel Fearing of his property. This money was a depreciated currency at the time and one wonders what real value Isaac Bumpus obtained when old tenner was later converted into new tenner, and this money depreciated to an even lower stage. Possibly Isaac was shrewd enough to convert this money into Spanish dollars.

Isaac Bump, the miller, was a grandson of Edward Bompasse, the original Bump settler. Isaac was also the twin of Zoar who moved to Virginia in his early manhood, with his brother Samuel. The Bumpus family scattered to other communities, some to Brockton and other Massachusetts cities. There were a lot of Bumps, male and female, and they spread out over the land, intermarrying until they were related to scores of other families.

Only one or two married into the Fearing families, as can be seen in the list of marriages included in this chronicle of events.

One Bump marriage of special note was the marriage of Rhoda Bump to Edward Sparrow, February 21, 1765. Edward Sparrow marched with Israel Fearing's Minute Men Company on the 19th of April 1775. Sparrow was eventually elevated to Colonel, the second highest ranking officer from Wareham. After the Revolutionary War, Colonel Sparrow moved to Middleboro with his family.

Lands of Bumpus
(From the Rochester Proprietor's Records)

"John Bumpus thrown the 3 lots by the sea to 40 acors on the west side of Woowoantit river where his home now stands Nov. the 1696."

"Jacob Bompey hath laid out to him 40 acors of land lying near Woodoantit river, this land lying noar the north end of his brother John's lott, brought to record January 16, 1698."

"John Bumpus hath laid out to him 80 acors of land adjoining to his bounded at the north by a rod."

John Bumpus bought 20 acres more December 1697.

John Bumpus from Seth Pope 40 acres.

John Bumpus hath 60 acres adjoining his land.

"Jacob Bumpus 25 acres E. Side of Weweantic part of land granted to Thomas Clarke shares June 1, 1707."

"John Bumpus 34 acres on Weweantit river, some salt meadow and swamp. One boxberry swamp on one side to a little brook on the other side and on one side by John Sherman land."

"Daniel Bumpus March 24, 1759 in Wareham bounded by land of Deacon Wilbur in range of Jedediah Bumpus on southerly side of County Road.

Nathaniel Bumpus 13 acres in Wareham.

Jacob Bumpus and Samuel Primas agreed to divide a lot near Primas' house January 1, 1708.

Item

James Sullivan Bump's daughter is the most famous of the Bump family, the Countess Magri who was Mrs. Tom Thumb. She was first known as Mercy Lavinia Warren Bump, born in Warrington of Middleboro October 3, 1841. After her first husband's death she married Count Magri. Her Warren line can be traced back to William the Conqueror of England.

The Bumpus families, who were so active in the early history of Wareham, were descended from an early emmigrant named Edward Bompasse who, it is said, was a Frenchman who earlier migrated to England, then emmigrated to Plymouth Colony on the ship *Fortune* in 1621. He first lived in Duxbury, but soon moved to Marshfield where eight of his known children were born.

Sarah .... born March 9, 1631
Elizabeth .... born March 9, 1633
*John .... born June 2, 1636
Edward .... born April 15, 1638
Joseph .... born February 15, 1639-40

Isaac .... born March 31, 1642
*Jacob .... born March 25, 1644
Hannah .... born April 3, 1646
Phillip .... no dates
Thomas .... no dates

Of all these children John and Isaac are of the most interest to us, since all of the Wareham Bumpus, Bumps, Bompos etc. are descended from these two men's marriages and subsequent children. They moved finally to Rochester in the section now known as Wareham. King Phillip's war caused the Bumpus families to move out of Middleboro to Wareham.

The land they owned in Scituate they sold, and purchased large tracts of land in old Rochester. From this land they cut timber and converted it in their own sawmill to lumber from which they built many houses.

A few of these houses remain today, two of significant historical interest: the property owned by James R. Lincoln, M.D. on Lincoln Hill, North Main Street, and the Fearing Tavern, restored by the Wareham Historical Society and kept as a museum. This building, sitting on a small lot of land, faces south and is open to the public through July and August of each year.

John Bump (brother of Isaac) married Sarah (Hunter) Bump, and these children were born to them in Middleboro:

Mary .... born 1671
John .... born 1673
Samuel .... born 1676

These children were born in Scituate:

James .... baptised June 16, 1678
Isaac (twin) .... baptised March 27, 1681
Zoar (twin) .... baptized March 27, 1681
Job .... baptised April 29, 1683

These children of John and Sarah Bump were born in Wareham, part of Rochester:

Sarah .... born September 16, 1685
Edward .... born September 16, 1688
Jeremiah .... born August 24, 1692

In their early manhood, Edward, Jeremiah and John Bump owned parts in the four Rochester Sea Lots with Isaac Bump the miller. Edward built a house on his land (still standing) overlooking the Wankinco River (now facing a shopping center).

These men are mentioned several times in the town's history

as officials holding office as Moderator, selectmen, constables, fence viewers and other offices.

The influence of the Bumpus family waned and died with the sale of Isaac's property to Israel Fearing. This vigorous branch of the Bump family was old and dying off and an era was passing.

Edward Bump and his wife Martha had these children:

Daniel .... born May 1, 1719
Rachel .... born April 20, 1720
Selathiel .... born May 31, 1722
Joanna .... born March 17, 1724
Edward .... born March 22, 1726
Elener .... born May 8, 1729
Lose (Louise) .... born October 6, 1731
Jedidah .... born November 25, 1732

Jonathan Bumpus and wife Mary's children:

Jorusha, daughter .... born May 1, 1719
Jonathan, son .... born March 21, 1721
Noah, son .... born March 27, 1724
Zorish (3rd daughter) .... born April 4-5, 1726

Benjamin Bumpus, wife Susanna:

Elizabeth .... born April 20, 1719
Philip .... born August 27, 1723
Anna .... born January 3, 1725
Rhoda .... born November 30, 1729
Mercy .... born September 14, 1735

Benjamin Bumpus, wife Ruth:

Elizabeth .... born March 24, 1757

Benjamin Bumpus, wife Abigail:

Elisha B. .... born September 1, 1815
Albert .... born October 20, 1817
Benjamin .... born November 22, 1819

Benjamin B. Bumpus, wife Zelphie:

Lydia .... born May 6, 1812
Betsey .... born November 19, 1814
Owen Churbuck .... born February 6, 1817
Zelphic .... born February 6, 1817
Ann .... born March 24, 1822
Benjamin .... born August 15, 1825
Lysander Noble .... born September 5, 1827
Henry Franin .... born December 2, 1830
David .... born September 1, 1832

MISCELLANEOUS GENEOLOGY

Curtis Tobey's children:

Hope Brigge .... born Feb. 17, 1806
Joshua Briggs .... born Feb. 9, 1807
Moses Swift .... born Oct. 17, 1810
Seth Fish .... born June 29, 1814

Prince Burgess' children:

Stephen Cowell .... born Nov. 30, 1809
Martha Dexter .... born Feb. 12, 1825

Isaac Bumpus, wife Nancy:

Eliza Coffin .... born Mar. 8, 1812
William Brock .... born Apr. 8, 1814
Ruth Covel .... born June 8, 1816

Children of John Bourne, wife Polly:

Bedui Gibbs .... born Jan. 1, 1805
Samuel Savary .... born Oct. 5, 1806
William Leonard .... born Feb. 22, 1809
Deborah Tobey .... born Apr. 4, 1811
Ebenezer .... born Sep. 1, 1814
Mary Lucinda (twin) .... born Jan. 25, 1819
John Franklin (twin) .... born Jan. 25, 1819

Jacob and Elizabeth Blackmore children:

Benjamin .... born Nov. 13, 1677 Scituate
Jacob .... born July 6, 1680 Scituate
Iccabod .... born Sep. 10, 1685 Wareham

John and Jacob owned property in Scituate while they resided there. They both moved to that part of Rochester now Wareham, about 1684 and both bought land of the original Proprietors. Their sons, Benjamin, Isaac, and Jeremiah also continued to add to their holdings in Rochester Proprietary including the sea lots on the Wankinco River at the confluence of the Agawam River.

The Bump family acquired large tracts of land, both in Rochester and in Wareham, consisting of what was known as the five Bumpus lots long after the original six great lots of the Agawam Purchase.

On one of these lots Isaac Bump, the miller, built his house and his first grist mill on what was then known as Bumpus Brook, now known as Rose Brook coming down from a swamp in the north woods and draining southerly to the lower end of a bog area then turning easterly and entering the mill Holding Pond. Remains of this grist mill were evident less than seventy-five years ago, as reported by Daisy Lovell author of *Glimpses of Early Wareham,*

Wareham's only published history to date.

The Bumpus (Bump, Bumpas) family was expanding rapidly by the large number of children each marriage produced. Some of these people became famous for many reasons, some became wealthy, some entered politics holding responsible offices.

Some of the Bump men were very industrious and clever, some making sizeable fortunes while others, less fortunate, were still able to command large income from their business interests. Others remained poor all their lives.

Many of the early Bump families moved out of Wareham and Rochester to the south and west to Ohio because of poor economic conditions in Wareham.

Samuel and Zoar Bumpus, children of John and Sarah, second generation, migrated to Virginia. Samuel settled in Amelia County and had five children. A genealogy of this family has been published, according to Mr. Edmund Staples, from whom some of this history was taken in his research for Revolutionary War Soldiers.

The Bumpus family was directly involved in the early iron industry in Wareham through the daughters of several Bump families marrying the owners of some of the iron foundries and industries.

Mary Bump married Seth Ellis 1729
Ruth Bumpus married Ebenezer Holmes 1755
Cloe Bumpus married Thomas Whitten 1764

This information is again attributed to extensive research made by Mr. Edmund Staples, who married Mary Hammond, of an old established New England family extensively involved in cranberry bogs, formerly areas where bog iron was found. (see *Cannon Balls and Cranberries* by Fredericka A. Barrows.)

The early times were times of scarcity of many things, especially money and luxuries, except for the very rich. The working people and farmers, the artisans and builders would barter their labor or production in exchange for other goods. The English pound, shilling, and pence were still used in the early Seventeenth Century, being gradually replaced by a value placed on "old Ten" or "Tenner".

It was not easy for a farmer, which most of the people were, to buy extra land or build a ship, operate a store or engage in any kind of business for two very good reasons: (1) most of the early Colonists worked from sun up to sun down to just barely make a living; and (2) money, even if they had a product to sell, was extremely scare and transportation except by ship was almost non-existent.

Regardless of these handicaps John and Jacob Bumpus bought land from the early proprietors of Rochester and Wareham. They erected a saw mill on a part of the site now occupied by the Tremont Nail Works. They cut and hauled timber from the virgin woods they had purchased. They converted the oak and pine into timbers and boards to build houses. Many of these kind of houses, strongly built, still stand in Wareham.

Isaac Bump sold this property to Israel Fearing, February 7, 1749, five years before Israel's death, and Benjamin Fearing as an heir of Israel Fearing the father, inherited this portion of his estate. The next heir, William S. Fearing, passed it on to his son, William H. Fearing whose son, William H. Jr., passed it on to William H. Fearing whose daughter Flora May Fearing married Alexander Warr. She bought out the heirs for $2,200.00. The heirs consisted of:

Fannie J. Fearing
Isiah F. Shurtleff
F. Mildred Kirby
Arthur A. Kirby
Louise T. Soule
Lincoln A. Soule
Marion C. Koster
Charles W. Koster

Sherman T. Fearing (who was single)

APPENDIX

A Note on Genealogy Research

If you who read this book are either Bumpus or Fearing or of another family named here, and desire to trace your "roots" back to those given in this chronicle, you can do so by purchasing a book entitled *Searching Your Ancestors* as a beginning, to afford you some insight into the procedure. The author is Gilbert H. Doane.

You can ask those oldest in your family for information as to where your ancestors came from, search church records, town records and cemeteries. Historical Societies are also good sources of information. Some states have archives and State Historical Societies from which you may obtain additional information.

Some antiquarian book stores carry old geneological works listing your family name and biographical information.

Your local library and state library are also good sources. The Library of Congress in Washington, D.C. has 25,000 printed family histories and 80,000 local histories.

The Church of Latter-Day Saints in Salt Lake City will assist Mormons and non-Mormons alike. You will need to contact a local unit of the church, as they do not answer letters. The National Archives in Washington, D.C., has branches in Boston, New York, Philadelphia, Atlanta, Chicago, Kansas City, Fort Worth, Denver, San Francisco, Los Angeles, and Seattle.

Another source in Washington is the Library of the Daughters of the American Revolution. It has a 60,000 volume library.

Some major sources are the New England Historic Geneological Society, Boston. The National Geneological Society, Washington; the Newbury Library, Chicago; the New York Public Library; the Maryland Historical Society and the Enoc Pratt Library, Baltimore.

If you do not want to do the work yourself, contact the Board for Certification of Genealogists, 1307 New Hampshire Avenue, Washington, D.C. 20036 for information.

Fraternal organizations such as the Elks, Moose, Eagles, Knight of Columbus, Masons, Shriners, etc., can possibly help you.

Assessment and tax records may be a source of help. Records of the Armed Services can help.

If your ancestors have moved from the place of their birth, one of the places to look is the New England Genealogical Society. It is custodian of the Revolutionary taxes imposed on every adult male tax payer. This record constitutes the most comprehensive listing of tax payers at that time. These are located in the Old Customs House in Boston, Mass.

Checking tax rolls in the town of origin of your ancestors may show their appearance on (or disappearance from) the tax rolls or assessors rolls. Deeds in the county registry of deeds could lead to more information.

You could begin with old letters, receipts and other family papers kept by relatives, old ledgers of long discontinued businesses, as well as cash books preserved in local historical societies and by local historical buffs.

Finally, if you need a start on a very early ancestor, you may need access to foreign sources. The College of Arms in England is the best place for information if your ancestor was of English extraction.

If your ancestor was from another foreign country, contact the

Ambassador from that country to the United States or a local Consul if there is one in your city.

Other sources will reveal themselves to you as your search progresses. You will meet other amateur genealogists and officials representing genealogical societies who will gladly assist you.

Genealogical research is rewarding, but filled with many frustrations and dead ends. It takes an enormous amount of time and can be very expensive, whether you do it yourself or hire an expert genealogist to search for you. It is at present a very popular process and as such you will find replies to your requests slow, due to overloaded personnel who have thousands of similar requests to answer. By persistence and patient waiting you should be rewarded.

INDEX

CRAB COVE
WAUKINCO RIVER
TIMS IS
NARROWS
WHARVES
BROAD MARSH RIVER
QUASUET
COTTON FACTORY
SHIPYARD